Construction Business Management

Construction Business Management

John Schaufelberger
University of Washington

Prentice Hall
Upper Saddle River, New Jersey
Columbus, Ohio

Library of Congress Cataloging-in-Publication Data

Schaufelberger, John
 Construction business management / John Schaufelberger.—1st ed.
   p. cm.
 Includes bibliographical references and index.
 ISBN-13: 978-0-13-090786-8 (alk. paper)
 ISBN-10: 0-13-090786-3 (alk. paper)
  1. Construction industry—Management.   I. Title.
 HD9715.A2S358 2009
 624.068—dc22

                                                2008021656

Editor in Chief: Vernon R. Anthony
Editor: Eric Krassow
Editorial Assistant: Sonya Kottcamp
Director of Marketing: David Gesell
Marketing Manager: Jimmy Stephens
Marketing Assistant: Les Roberts
Production Manager: Kathy Sleys
Creative Director: Jayne Conte
Cover Design: Bruce Kenselaar
Cover Illustration/Photo: © M. Dillon/CORBIS. All rights reserved
Full-Service Project Management/Composition: Shiny Rajesh/Integra Software Services Pvt. Ltd.
Printer/Binder: Hamilton Printing Company
Cover Printer: Phoenix Color Corp.

Credits and acknowledgments borrowed from other sources and reproduced, with permission, in this textbook appear on appropriate page within text.
The names of companies and their employees used in the examples in this book are fictional, and are not associated in any way with any actual companies or persons that may share the same names(s).

Copyright © 2009 by Pearson Education, Inc., Upper Saddle River, New Jersey, 07458.
All rights reserved. Printed in the United States of America. This publication is protected by Copyright and permission should be obtained from the publisher prior to any prohibited reproduction, storage in a retrieval system, or transmission in any form or by any means, electronic, mechanical, photocopying, recording, or likewise. For information regarding permission(s), write to: Rights and Permissions Department.

Pearson Education Ltd., London
Pearson Education Singapore, Pte. Ltd
Pearson Education Canada, Inc.
Pearson Education–Japan
Pearson Education Australia PTY, Limited

Pearson Education North Asia, Ltd., Hong Kong
Pearson Educación de Mexico, S.A. de C.V.
Pearson Education Malaysia, Pte. Ltd.
Pearson Education Upper Saddle River, New Jersey

Prentice Hall is an imprint of

PEARSON

www.pearsonhighered.com

10 9 8 7 6 5 4 3
ISBN-13: 978-0-13-090786-8
ISBN-10:   0-13-090786-3

Contents

Preface   x

CHAPTER 1   Introduction   1

Nature of Construction Business   1

Primary Causes of Business Failure   2
  External Influences   2
  Internal Problems   2

Business Strategies to Minimize the Risk of Business Failure   4
  Planning Strategies   4
  Plan Implementation/Control Strategies   5
  Financial Management Strategies   6

Leadership Challenges   7

Organizational Behavior   8
  Employee Motivation   8
  Individual Behavior   9
  Collective Behavior   10

Ethics   11

Summary   13

Review Questions   14

Exercises   14

Sources of Additional Information   15

CHAPTER 2   Company Organization   16

Introduction   16

Alternative Forms of Business Organization   16
  Proprietorships   17
  General Partnerships   18
  Limited Partnerships   19
  Corporations   20
  Type S Corporations   21
  Limited Liability Companies   22
  Joint Ventures   22

Alternative Organizational Structures   23
  Functional Organization Structure   24
  Divisional Organizational Structure   25
  Matrix Organizational Structure   26

Organizational Design   27

Policies and Operating Procedures   30

Summary   32

Review Questions   32

Exercises   32

Sources of Additional Information   33

CHAPTER 3   Risk Management   34

Introduction   34

Risk Management Strategies   35

Insurance   36
Insurance Contracts   36
Insurance Agents   37
Insurance Types   38
Subcontractor Insurance   48
Insurance Certificate   48

Bonding   48
Surety Relationship   48
Bonding Agents   52
Types of Bonds   52
Underwriting Considerations   58

Summary   64

Review Questions   65

Exercises   66

Sources of Additional Information   66

CHAPTER 4   Financial Analysis and Management   67

Introduction   67

Accounting Systems   67
Primary Accounting System   68
Secondary Accounting Systems   69
Accounting Records   71

Accounting Methods   72

Financial Statements   73
Balance Sheet   73
Income Statement   75

Financial Analysis   76
Financial Statement Analysis   76
Cash Flow Analysis   81

Financial Management   84

Summary   88

Review Questions   89

Exercises   89

Sources of Additional Information   90

CHAPTER 5 Strategic Planning and Management 91

Introduction   91

Planning Process   93
  Mission Statement   95
  Company Vision   96
  Strategic Assessment   97
  Strategic Objectives   97
  Company Strategies and Short-Term Goals   98
  Action Plans   98
  Performance Measurement   99

Situation Analysis   99
  External Analysis   99
  Internal Analysis   105

Strategy Formulation   110

Strategy Implementation   113

Strategy Evaluation   113

Summary   114

Review Questions   115

Exercises   116

Sources of Additional Information   116

CHAPTER 6 Business Development 117

Introduction   117

Marketing Construction Services   119

Marketing Process   120

Market Analysis   122
  Demand Assessment   123
  Customer Satisfaction Assessment   124
  Competition Assessment   125

Marketing Strategies   126

Marketing Tools   129

Marketing Plan   131

Acquisition of Work   133

Summary   133

Review Questions   134

Exercises   134

Sources of Additional Information   135

CHAPTER 7 Human Resources Management 136

Introduction   136

The Challenge   136

Decision-Making Processes   138

Company Culture 138
Organizational Design 140
Staffing 142
Employment Manual 147
Employee Development 149
Leadership Development 152
Succession Planning 153
Performance Management 154
Compensation and Employee Benefits 156
Employee Retention 156
Union Relations 157
Safety and Wellness Programs 158
Regulatory Overview 160
Affirmative Action 160
Age Discrimination 161
Disabled Workers 161
Discrimination 161
Drugs in the Workplace 162
Equal Employment Opportunity 162
Preventing Violence in the Workplace 163
Privacy Rights 163
Racial and Ethnic Discrimination 163
Record Keeping 164
Religious Protection 164
Sexual Harassment 165
Summary 166
Review Questions 166
Exercises 167
Sources of Additional Information 167

CHAPTER 8 Information Management 168

Introduction 168
Information Requirements 170
Hardware and Infrastructure 171
Software 172
Security Systems 173
Company Security Policy 174
Company Web Sites 174
Summary 175
Review Questions 175
Exercises 176
Sources of Additional Material 176

CHAPTER 9 Total Quality Management   177

    Introduction   177

    Quality Concepts   179
        Quality Principles   180
        Supporting Elements   181
        Process Improvement Model   182

    Implementation   183

    Process Analysis   187

    Sustainment   188

    Effectiveness Assessment   191

    Summary   191

    Review Questions   192

    Exercises   192

    Sources of Additional Information   193

Appendixes

    A—Strategic Plan for Pacific Constructors   194

    B—Case Study—Cascade Builders Annual Report   200

    C—Case Study—Northwest Constructors Annual Report   209

    D—Employment Manual for Western States Construction Company   224

    Glossary   242

    Index   247

Preface

Construction is a risky business, and construction company managers often lack the necessary business management skills needed to ensure the survival of their firms. While good cost-estimating, project planning and scheduling, and project management skills are essential for success in construction, so are good business planning and management skills. Unlike other industries, the construction industry has few executive development programs, but relies primarily on on-the-job training. Most construction company leaders learn to run a business by watching others or by working with people who have been successful in managing construction firms. Many of these leaders lack formal business management education. Often they were promoted from project management positions within their companies or formed their own firms after working for others.

The business failure rate in construction is about 30 percent higher than the national average for all industries. This statistic provides a measure of the risk faced by construction company leaders and highlights the need for good business management skills. What are the reasons why construction firms get into financial difficulty? Some are external, such as economic downturns, but construction company leaders must understand that construction can be cyclical and that they must develop contingency plans for changes in the market. Many reasons, however, are internal and within the leader's ability to control. These may be grouped into the seven categories discussed below.

Pursuit of volume. Volume, not profitability, is often used as a measure of a contractor's success. Rarely are profit margins, return on equity, and changes in profitability used to describe the condition of a construction firm. This constant pursuit of volume may lead to accepting lower profit margins on projects because of intense competition. Rapid growth also stresses the firm's management systems and may spread management too thin to adequately monitor the performance of individual projects.

Lack of comprehensive business plans. Some construction firms do not have business plans that guide their business decisions. They simply react to the market. Business planning requires an understanding of the market, how construction procurement decisions are made, and the competitive advantages of the firm. This knowledge is essential in the selection of services to be offered, the selection of market area and focus, and the selection of people and equipment. In addition to understanding the market, the construction company leader must understand the firm's financial condition and devise strategies for financial success in both expanding and contracting markets.

Ineffective financial management. Contractors may not understand their firms' cost structure, which may result from accounting systems that do not provide the needed detail of cost data. In other cases, they may not manage their cash flow requirements properly or are not adequately capitalized and must resort to unplanned borrowing of capital. Another problem is the use of working capital to finance equipment purchases, thereby reducing the firm's ability to finance its cash flow requirements.

Poor internal communications. Poor internal communications between project sites and the home office plague many construction companies. Consequently, there may be little warning of project execution problems or financial difficulties. In most cases, it takes only one or two disastrous projects to bring down a company. Early warning is essential if corrective action is to be taken on time.

Inadequate marketing. Many contractors do little marketing, but focus on selling their services. Marketing involves everything a company does to retain and attract customers. Hard selling is of little value unless the firm has a reputation for quality customer service.

Poor human resources management. Construction firms succeed or fail based on the quality, skills, and motivation of their employees. Often this critical function is overlooked by construction company leaders. Quality team members must be recruited and given the skills needed to provide excellent customer service. They must be treated properly and recognized for their contributions to the company's success.

Unplanned leadership changes. Unplanned changes in leadership can be devastating to a construction firm, whether it be from retirement, illness, death, or resignation. Succession planning is essential to ensure continuity of operations. This is particularly important in family-owned businesses.

To be successful, construction company leaders must realize that their firms will survive only if the companies provide value to their customers and that customer value is provided with superior business practices, skilled people delivering quality services, and a good working environment to motivate their employees. Positioning their companies for success requires careful planning and periodic evaluation to ensure that responsible individuals have been given adequate resources to achieve desired goals and objectives. Management decisions made by company leaders will determine whether their firms succeed or fail.

This book was written to address the basic business skills that a construction company leader needs in order to be successful in the industry. It was written for use as a text in undergraduate and graduate construction management programs and as a reference for construction professionals. Most construction curriculums address the technical aspects of managing a construction project adequately, but few address the challenges of managing a construction company. My purpose in writing this book was to create a resource for construction educators and for construction firm leaders.

The book is organized into individual chapters that address the major business management functional areas. Each chapter concludes with a set of review questions that emphasize the major points covered in the chapter. Exercises are provided that require application of the principles discussed. The chapters have a list of other publications for those interested in additional information on the topics addressed. A sample strategic plan for a hypothetical construction firm is in Appendix A. There are two case studies in the appendix. The one in Appendix B is for a hypothetical residential construction company, and the one in Appendix C is for a hypothetical commercial construction company. Appendix D contains a sample employment manual for a construction firm. A glossary of terms is provided at the end of the book. An instructor's manual containing answers to the review questions and exercises is available.

To access supplementary materials online, instructors need to request an instructor access code. Go to **www.pearsonhighered.com/irc**, where you can register for an instructor access code. Within 48 hours after registering, you will receive a confirming e-mail, including an instructor access code. Once you have received your code, go to the site and log on for full instructions on downloading the materials you wish to use.

This book could not have been written without the help of many people. I wish to thank my students who used draft versions of the text and provided suggested improvements, the Prentice-Hall staff for their outstanding support, and the following reviewers for their helpful comments: Marjorie Parry Callahan, University of Oklahoma; Farid Jean Sabongi, Minnesota State University; Matt Syal, Michigan State University; and Kenneth J. Tiss, The State University of New York.

JOHN SCHAUFELBERGER
University of Washington

CHAPTER 1
Introduction

NATURE OF CONSTRUCTION BUSINESS

Construction involves the marshaling of materials, people, and equipment on a project site and assembling the materials in the proper sequence to construct a project that meets the customer's requirements. These projects may range from an individual home to a sophisticated infrastructure project, such as a regional airport or a major transportation system. The business management challenges in construction are to ensure that

- the revenue generated by the construction activity exceeds the cost of doing the work,
- the company has adequate demand for its services,
- the company has adequate financial resources to finance construction projects until reimbursed by its customers,
- the company has a skilled, motivated workforce of sufficient size to meet anticipated requirements, and
- the cost of the company overhead is affordable based on the projected workload.

Construction is an intensely competitive industry, with companies ranging in size from less than ten employees to over tens of thousands of employees. Because of the great diversity in the types and sizes of projects as well as the variety in the expertise and size of companies, most firms tend to specialize in distinct segments of the market, such as highway, commercial, industrial, residential, electrical, mechanical, site and utility, marine, and underground. Each segment of the market has its unique set of technical challenges, but the following business responsibilities are similar:

- acquisition of work,
- performance of the work, and
- management of the financial, capital, and human resources of the firm.

Construction is a risky business. About half of the construction firms in the United States fail before they complete ten years in business. Only Internet companies and the food service industry have a higher bankruptcy rate than does the construction industry. The construction industry annually accounts for about 10 percent of the gross domestic product of the United States. It is a fragmented industry with a large number of firms. Unlike manufacturing, construction does not require large capital investments to establish a business. Construction, however, does require sufficient cash resources to meet financial obligations.

CHAPTER 1   Introduction

In this chapter, we will discuss some of the primary reasons construction firms fail and some strategies that will minimize the potential for business failure. Then we will discuss some of the leadership challenges faced by construction company leaders. The next topic that we will discuss is some basic concepts of organizational behavior and how they can be applied to improve performance of a construction company. Since the business of a company is conducted by its employees, it is important that company leaders understand how people respond as members of an organization. The last topic that we will discuss is company ethics and the role ethics plays in establishing a company's reputation.

PRIMARY CAUSES OF BUSINESS FAILURE

Few construction firms fail from a single cause or from a sudden, catastrophic event. One cause may predominate, but inadequate response to several interrelated factors is the typical cause of business failure. The primary causes of failure can be grouped into two categories: external influences and internal problems.

External Influences

The major external influences on business failure are

- prolonged economic recession,
- loss of a major customer,
- new competition, and
- shortage of skilled labor.

Recessions result in reduced construction activity, which drives more construction companies to compete for that work which is available. An oversupply of contractors pursuing limited work often leads to reduced profit margins, jeopardizing the economic viability of winning construction firms. A construction firm's major customer may significantly reduce its construction activity or even may go out of business. If the construction company does not have other customers, the loss of a major customer may cause severe financial strain on the company. New competition also presents threats to the construction firm, particularly if the size of the market is not increasing. Intense competition often results in lower profit margins and greater risk. Without an adequate supply of skilled craftsmen, a construction firm may be unable to produce quality work desired by its customers, resulting in the loss of reputation and business.

Internal Problems

Despite the external threats discussed above, the primary causes of construction firm failure are the result of one or more internal problems. These are discussed below.

Strategic Planning Issues

Pursuit of volume   To many people in the construction industry, business volume is a measure of success. While it may indicate the significance of a firm in its relevant market, profitability is a more important measure of success. The pursuit of volume without a corresponding increase in profitability places the economic viability of the construction firm at risk.

Lack of comprehensive business plan   Construction companies often do not have business plans that guide their business decisions. They may simply react to the market. Business planning requires an understanding of the market, of how construction procurement decisions are made, and of the competitive advantages of the firm. This knowledge is essential in the selection of services to be offered, the selection of market area and focus, and the selection of people and equipment required. In addition to understanding the market, construction company leaders must understand their firms' financial condition and devise strategies for financial success.

Diversifying into unfamiliar types of projects   There is high risk in diversifying into unfamiliar types of projects, because appropriate suppliers and subcontractors may not be known, and the technical requirements may exceed the expertise of the company's project management staff. Costs may be underestimated, resulting in unprofitable projects. Similarly, pursuing projects using unfamiliar contracting approaches, such as design-build, often results in significant financial risk to the construction company.

Diversifying into unfamiliar geographic areas   Entering an unfamiliar construction market poses great risk to a construction firm. Potential customers, suppliers, and subcontractors are unknown. If the location of the new market is a significant distance from the company's normal area of operation, there will be little ability to augment project management staff with other company resources; placing greater stress on the project management team.

Lack of managerial maturity   Construction firms often are founded by one or two people. As a firm grows, its management system must mature to accommodate increased scope of work. Additional managers are needed, and the founders must delegate some authority for making business decisions to others. Continuing to centralize all decision making in one or two people does not provide the responsiveness needed to react to changing business conditions.

Strategic Implementation/Control Issues

Increase in project size   Unrealistic increase in project size may lead to financial difficulties. The size of the project in relation to the capabilities of the firm may lead to significant financial problems. One large unprofitable project will have a greater adverse impact than will a small unprofitable project. Large projects may stress managerial expertise, subject the firm to greater risk, and require more capital to finance cash flow requirements.

Unplanned loss of key personnel   Unplanned loss of key personnel can severely stress construction firms, particularly small ones. The loss can be due to death, to illness, or to resignation. Unplanned loss of management or technical expertise may take considerable time to overcome and places small firms at great risk.

Poor cost-estimating skills   To ensure that a company remains a profitable business enterprise, construction firm managers must understand the anticipated costs and risks of each project and price the work at a level to cover all costs and provide a profit.

Because of some adverse condition, the project may not provide the anticipated profit, but construction company managers should not pursue volume at the expense of profitability. This means that bids and cost proposals should not be reduced just to increase the company workload, because with greater work volume the company incurs more risk.

Lack of equipment control   The cost of owning and operating equipment is a significant part of the construction business. Controlling equipment costs is controlling the amount of equipment owned, leased, or rented. Investing too much of a company's financial resources in equipment may degrade the firm's ability to finance its cash flow requirements. Idle equipment represents an overhead cost that should be avoided. Equipment should either be used and be able to pay for itself, or returned to the rental or leasing company or sold.

Poor internal communications   Poor internal communications between project sites and the home office plague many construction companies. Consequently, there may be little warning of project execution problems or financial difficulties. In most instances, it only takes one or two disastrous projects to bring down a company. Early warning is essential if corrective action is to be taken on time.

Financial Management Issues

Poor use of accounting systems   A number of contractor failures are caused by poor accounting practices or by a failure to review accounting records to determine the financial status of projects. A delayed customer payment often results in inadequate cash flow. By not using the current balance sheet, income statement, and job cost reports, managers may be unable to identify financial difficulties until it is too late to take corrective action.

Excessive debt   Cash flow requirements must be met either from internal resources or from debt. Some contractors rely on debt to compensate for the lack of business capitalization or equity. Excessive debt may cause such a drain on income that the capital resources of the firm do not grow or may even decline. The cost of borrowing should be included in company overhead margins. Too high margins may make the construction company noncompetitive in tight markets.

BUSINESS STRATEGIES TO MINIMIZE THE RISK OF BUSINESS FAILURE

To minimize the potential for business failure, construction company owners and leaders should adopt the following strategies:

Planning Strategies

Develop comprehensive business plans and communicate company goals and objectives to all employees. In order for a company to be successful, all employees must understand the company values, where the company leaders intend to take the firm, and the employees' roles in making the company successful. The business plan provides a tool for assessing the company's current position and for establishing a vision of what the company aspires to become. In addition to containing goals to be accomplished, the plan should identify specific action plans that are to be undertaken by different segments of the company to assist in meeting the business goals.

Test a new geographic area with a small project or a repeat customer. Develop a withdrawal plan if the project is not successful. Moving into a new geographic area presents significant risk, in that the customers, the suppliers, and the subcontractors generally are unknown. To reduce the risk, either undertake a series of small projects to learn the environment or enter the market with a customer with whom the company is familiar.

Carefully monitor company management capabilities during periods of growth. Learn to recognize signs of inadequate management, and delegate. Adequate company management systems and staff need to be available to manage the work of the company. Management capabilities need to be expanded prior to increasing the volume of work undertaken, so that the increased work can be planned and managed effectively. Sometimes companies increase their work volume without adequately increasing the management capabilities. In such instances, company managers become overworked and may choose to seek employment elsewhere. Also, the lack of adequate management capability may result in poor project performance and an unhappy customer, which is undesirable.

Plan Implementation/Control Strategies

Develop internal management systems that monitor the status of all projects and provide early warning of problem areas. Good communications need to be established between all project offices and the company leadership. In addition, company leaders need to visit project sites frequently to keep abreast of any issues. Frequent project cost and project status reports are needed to closely monitor project execution.

Form long-term relationships with industry professionals who can serve as sounding boards and listen to them. Among these specialists are bankers, bonding agents, insurance agents, accountants, and construction attorneys. Because construction is a specialized business requiring support from a variety of industry professionals, company leaders should establish enduring relationships with a banker, a bonding agent, an insurance agent, an accountant, and an attorney. These individuals should be selected based on their industry experience and understanding of the type of projects undertaken by the construction company. Because they have industry experience, they can provide advice regarding project risk mitigation and business processes.

Increase project size gradually. Take on only one larger project at a time. Finish the first larger project and evaluate before taking on the next one. The amount of risk associated with a project generally relates to its size and its complexity. Care must be exercised when pursuing projects that are larger than those historically constructed by the firm. If the typical project size has been $5 million, it is better to pursue a project valued at $7 million than one at $15 million. This is to ensure that the company's management expertise and systems are enhanced to meet the demands of larger projects. Undertaking too large a project usually requires more cash flow than the company should risk on a single project.

Select project size based on the size of the construction company and its financial and human resources. Do not have more than 30 percent of a company's resources involved in a single project. Diversify among several projects to minimize risk. Success in the construction business depends on understanding the risk faced on each project and developing plans for managing the risks presented. Not all projects are successful. Sometimes things do not work out as anticipated, and the construction firm incurs a loss. To minimize the potential for financial failure if a project is not profitable, the construction firm should undertake multiple projects, as long as it has sufficient managerial talent to supervise the projects effectively.

Financial Management Strategies

Understand the cost of doing business, and price services appropriately. It is essential that company leaders understand their cost of doing business to include all company overhead costs. Services must be priced adequately to cover all anticipated costs and provide a reasonable profit. If the company does not adequately price its services, it will soon find that it is out of cash and in financial difficulty.

Prepare month-to-month cash flow budgets each year, and track results monthly. Company leaders need to understand their financial needs to ensure that they have the ability to meet financial obligations. In most cases, construction work is initially financed by a construction company, and a bill or invoice is submitted to the project owner for payment. This request for payment typically is submitted on each project on a monthly basis. A cash flow analysis to develop a cash flow budget for the construction company is needed for each project. If external financing is needed, the cost of the financing needs to be included as an overhead cost while calculating project budgets. Actual income and expenditures need to be tracked to monitor the cash flow status of the company.

Carefully manage company overhead, and reduce it during periods of declining workload. Company overhead budgets need to be set at the beginning of each year based on what is perceived as affordable for a projected volume of work. If the volume of work does not materialize, the overhead budget needs to be reduced proportionately so that the company does not get into financial difficulty because company earnings were insufficient to cover overhead costs.

Finance equipment purchases with debt to preserve capital to finance cash flow requirements. Capital equipment should be purchased using either a lease-to-own strategy or by the use of equipment loans. The concept is for each item of equipment to earn more each year than it would cost to make a loan payment or a lease payment. This will preserve the construction company's financial resources to fund its cash flow requirements. If the cash is used to purchase equipment, the company could become cash-starved relative to meeting its cash flow requirements and forced to go out of business.

Purchase a line of credit as a contingency to cover unexpected negative cash flows. Rather than waiting until there is a need to borrow funds, a company should purchase a line of credit from a financial institution. This is similar to a credit card, in that expenses can be charged to the line of credit when needed, and the amount of credit can be repaid as soon as sufficient income is received. There is a small fee for having the line of credit, but interest is only paid on the unpaid balance at the end of each month, similar to the situation with a personal credit card.

LEADERSHIP CHALLENGES

The chief executive officer or owner of a construction firm is the focal point that everyone within the company looks at for direction, leadership, values, and recognition and assurance that the company is moving in the proper direction. Company leaders, knowingly or not, form their companies in their own images. They set the standards for company ethics and values by their actions. These leaders must understand that the success of their companies lies in their ability to meet the needs and expectations of both their employees and their customers. Company leaders need to listen to their firms' employees and customers and then formulate goals, methods, and directions for their companies.

Construction company leaders should also be good managers. The basic management functions that they perform are planning, organizing, staffing, leading, and controlling. **Planning** includes defining company goals, establishing strategies to accomplish goals, and developing plans to coordinate the activities of organizational elements. **Organizing** is determining what tasks are to be done, who is to do them, how the tasks are to be grouped, who reports to whom, and where decisions are made. It also involves the selection of individuals to occupy key company positions. **Staffing** includes the recruitment, selection, and retention of company employees. **Leading** includes creating a vision for the company, motivating subordinates, establishing company standards of behavior and ethics, directing others, selecting communication channels, and resolving conflicts. **Controlling** is monitoring activities to ensure that they are being accomplished as planned and correcting any significant deviations.

The basic roles of a company leader have been described as the following:[1]

- Interpersonal role based on the formal authority defined by the company organization chart.
- Informational role as a recipient and transmitter of information.
- Decisional role as a decision maker.

Company leaders need to understand these roles and how their actions influence the attitudes, values, and performances of others within their companies.

Leadership is the process of creating a vision of what the organization is to become and having the ability to translate that vision into reality and sustain it. A good leader needs the following:

- Ability to create a vision and pull people forward toward its accomplishment.
- Ability to communicate effectively.
- Ability to empower subordinates by sharing power with them.
- Ability to recognize their own strengths and weaknesses.

The vision guides the decisions of company leaders and aligns the work of employees so that they work together in the pursuit of common goals.

The effectiveness of the leadership of a company can be measured in three ways: the company's business results, its organizational structure, and its culture. Company leaders affect the business result of their firms not only through changes in the organizational structure, but also by the culture that they create. They establish the company culture by identifying and communicating core values, specifying expected employee

---

[1]Mintzberg, Henry. *The Nature of Managerial Work*, Englewood Cliffs, N.J.: Prentice-Hall, 1980, pp. 91–92.

behaviors, establishing methods for providing performance feedback to employees, and creating supporting recognition and reward systems. Effective leaders should be ethical, fair, honest, goal-oriented, decisive, consistent, and reliable.

Many successful construction company senior executives devote their efforts to strategic planning, employee selection and development, and customer interaction. They leave the day-to-day management decisions to mid-level leaders. This approach serves two very important functions for a construction firm. First, it frees the senior executives to focus on long-range planning issues. Second, it provides professional development opportunities for the mid-level leaders and demonstrates the trust placed in them. To be able to delegate day-to-day decision making, the company leaders need to ensure that the organizational structure of the company supports decentralized decision making and that appropriate management processes have been implemented. While decision-making authority can be delegated, the ultimate responsibility for the well-being of the firm still rests with the senior leaders.

Construction company leaders need to understand that their companies have two sets of customers, internal and external. The **external customers** are those firms and individuals that purchase construction services from the construction company. The **internal customers** are the employees who receive support from various departments within the company. The company's employees represent the company's core capabilities and need to be nurtured in a manner similar to the company's external customers. They are the human capital the firm needs to be able to provide quality services to its external customers. Good leaders have a sincere and obvious caring for their subordinates and are willing to invest in employee training and development. These leaders align their talent development strategies with the strategic plans of their companies.

Company leaders get things accomplished through the members of their company teams. These leaders make decisions, allocate resources, and motivate others to accomplish company goals. They should possess four types of skills to be successful. They must have the *technical skills* to understand the technical aspects of the construction business. They must possess the *people skills* needed to work with, understand, and motivate people, both individually and in groups. They must have the *communicative skills* to be able to communicate effectively with employees, customers, and the media. Lastly, they should have the *conceptual skills* needed to be able to analyze complex situations and identify alternative solutions.

ORGANIZATIONAL BEHAVIOR

To be effective, construction firm leaders must understand how people behave in organizations and how to motivate people to perform and to want to remain as members of the company team. The primary measures of employee motivation are job performance, absenteeism, and turnover.

Employee Motivation

The quality of job performance is influenced by

- an employee's understanding of the importance of the job to company success,
- the employee's perception of how he or she is valued within the company organization,

- the employee's compensation,
- whether or not he or she is publicly recognized for the quality of work performed, and
- the working environment.

The work environment has two major aspects. One is the physical environment, and the other is the interpersonal environment, which relates to the interpersonal behavior of the company staff. Absenteeism generally is a function of an employee's satisfaction with the work environment, job responsibilities, and recognition within the company. Turnover is influenced by the same factors as absenteeism as well as compensation and the availability of opportunities for development and advancement.

Company leaders face motivational challenges at two levels: at the individual level and at the company level. They must focus on motivating each person as well as on creating motivating cultures and environments within their companies that encourage employees to work together as collaborative teams.

Social scientists have argued that human behavior is based on needs. One of the more well-known theories was developed by Abraham Maslow.[2] He categorized human need in the following hierarchical levels:

Higher-order needs

- *Self-actualization*—needs related to personal growth, self-fulfillment, and realization of one's potential.
- *Esteem*—need for self-respect, prestige, and recognition.
- *Social or affiliation*—need for love, belonging, and affection.

Lower-order needs

- *Safety or security*—need for security and protection from physical or emotional harm.
- *Physiological*—basic human needs such as food and water.

This hierarchy is illustrated in Figure 1.1. Maslow postulated that once a given level of need is satisfied, the next higher level of need is actuated. For example, once the basic need of food and water is received, an individual will be concerned with protection from harm. Understanding the relationship among the higher-order needs is essential for motivating individuals to want to be company employees and become high performers.

Understanding why people behave in the manner they do is important in developing strategies for employee motivation. A good leader recognizes that a higher level cannot be achieved until a previous level has been satisfied. Employees who perceive that their needs are not being met may not be motivated, may have poor job performance, and may choose to leave the company.

Individual Behavior

Construction companies are made up of unique individuals, who possess different personalities as well as different attitudes about a variety of things. Personality and attitudes are individual differences that employees bring to the workplace, which can

---

[2]Maslow, Abraham H. "A Theory of Human Motivation," *Psychological Review*, July 1943, pp. 370–396.

10 CHAPTER 1   Introduction

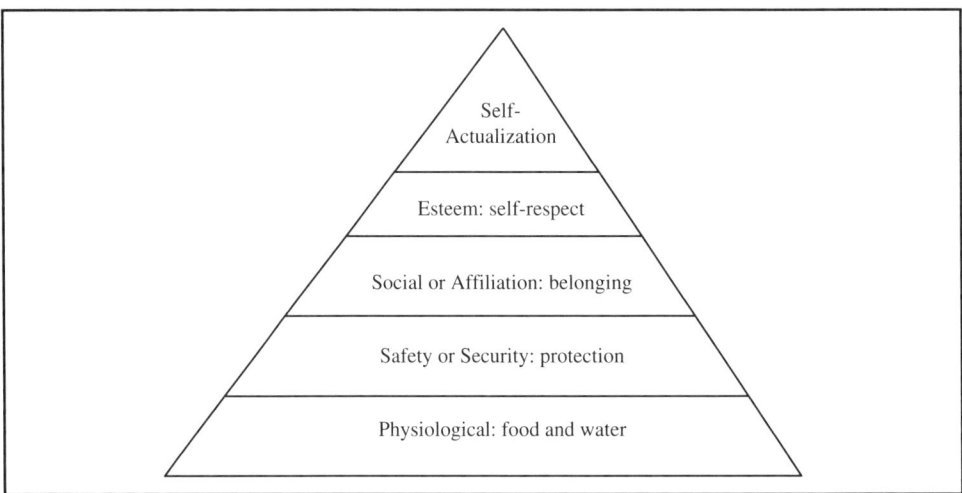

FIGURE 1.1 Maslow's Hierarchy of Needs

affect their behavior at work. An individual's personality directly influences job performance. Because one's personality is developed before being hired, managers need to determine which positions best fit each current and prospective employee's personality. Individuals whose personalities are not compatible with the job requirements of a position might not be successful in that position.

In general, there are five basic characteristics to an individual's personality. These are emotional stability, interpersonal behavior (extraversion or introversion), agreeableness, conscientiousness, and openness to change.[3] Attitude formation is primarily a learning process. Attitudes may be formed by personal experience, by education, or by observation. While it is difficult, if not impossible, to change an individual's personality, it is possible to change his or her attitudes toward specific issues. Attitude adjustment can be an important aspect of employee motivation. Key tools for individual motivation are ensuring that each employee understands the importance of his or her job to the accomplishment of company goals, treating each employee with respect, and publicly recognizing good performance.

Collective Behavior

The last aspect of organizational behavior to be considered is how people behave within an organization. Organizations function when members work together as teams to accomplish collective tasks. These may be formal assigned teams or they may be informal teams. Formal teams may be assigned as a part of the company's organizational structure, such as a project management team, or they may be ad hoc teams assigned to address specific issues, such as a team to plan a company social function. Informal teams are the informal contacts an employee establishes to address specific issues, such as preparing certain types of documents. The strength of an organization

---

[3]Barrick, M. R. and M. K. Mount. "The Big Five Personality Dimensions and Job Performance: A Meta-Analysis," *Personnel Psychology*, vol. 44, 1991, pp. 1–26.

comes from the synergy of people working together on either formal or informal teams to accomplish collective goals and objectives. Understanding how to motivate people to work together is an essential skill needed by construction firm leaders.

ETHICS

A company is judged by the integrity that it demonstrates in conducting its business and how it treats its employees. A key component of a company's reputation is the ethics of the company leaders and the company's employees. **Ethics** are the moral standards used by people in making personal and business decisions. We all face choices in our personal and professional lives that require decisions. The set of moral standards that we use to guide our decision making are our ethics, which help us to decide "what is right and what is wrong." An example of an ethics guide is the so-called Golden Rule, which is, to treat others as we would wish to be treated.

Many professional organizations have developed codes of ethical behavior, such as the one shown in Figure 1.2 developed by the Associated General Contractors of Washington. These codes of ethical behavior have been adopted based on the values of the members of the organization and are used to provide a framework for making ethical decisions.

Ethics involve determining what is right in a given situation, and then having the courage to do what is right. Each decision we make has consequences, often to us and to others. Generally there are considered three primary ethical directives: loyalty, honesty, and responsibility. Loyalty may be requested from many groups and institutions: friends, family, employer, profession, and society. Honesty is more than truth telling. It involves not lying, but, more importantly, involves the correct representation of ourselves, our actions, and our views. Responsibility means anticipating the potential consequences of our actions and taking responsible measures to prevent harmful occurrences.

FIGURE 1.2 Sample Code of Ethical Behavior

> **Associated General Contractors of Washington**
>
> **CODE OF ETHICAL CONDUCT**
>
> I believe that each of us, as individuals and organizations, owe our community, colleagues, customers, and each other a duty and responsibility to conduct ourselves in an honest and ethical manner.
>
> To further express my commitment to the principles of skill, responsibility, integrity, and community, I pledge to conduct myself according to the following standards:
>
> My word is my bond and is stronger than a written contract.
> I will treat others as I desire and expect to be treated by them.
> I will put safety and compliance with codes, laws, and regulations above profit.
> I will respect and protect the environment.
> I will do my best to produce quality projects on time, at good value.
> I will assume responsibility for my actions.
> I will strive to reach accord through personal negotiations and do my best to resolve disputes quickly, with integrity, and without personal attack or rancor.
> I will endeavor to persuade others within my organization and all for whom I am responsible to embrace these standards.

A construction firm's ability to acquire and maintain customers will be greatly influenced by the potential customers' perception of the ethics of the construction company and its employees. Company leaders establish the corporate culture regarding ethical behavior and establish the standards for their employees by their own personal actions. Subcontractors, material suppliers, and project owners may choose not to do business with general contractors who have a reputation for unethical behavior.

Construction company employees will face many ethical dilemmas in contract procurement, cost estimating, project management, accounting/financial management, customer relations, subcontractor relations, and vendor relations. Ethical behavior is a difficult area because it involves more than simply complying with legal requirements. It involves treating everyone in a responsible and fair manner. A company's reputation for ethical behavior may also affect its ability to attract needed talent. A company's ethics need to be one of its **core values** and stressed to all employees by company leaders. Ethical issues are very important, because they have the power to damage the image of the company and destroy the morale of its employees.

To create an ethical framework for a company, company leaders should create a company statement of values that will provide moral guidelines for company employees. A company statement of values should contain five to ten value statements that define the essential characteristics of appropriate employee behavior. An example statement of values is shown in Figure 1.3. Not only must the values be clear, but they must be reinforced by the daily activities of company leaders. Another tool for encouraging ethical behavior in company employees is a statement of business conduct, such as the one illustrated in Figure 1.4. This statement provides guidance to employees in making their daily business decisions.

FIGURE 1.3 Statement of Values for Western Construction*

We, the employees of Western Construction, are guided by the following values. They describe who we are and guide our decisions.

- *We take responsibility for the quality of our work.* We ensure that all of our work conforms to contract requirements.
- *We take responsibility for ensuring the safety of our work sites.* Our goal is no accidents on any of our work sites. We are committed to providing a safe working environment.
- *We take responsibility for providing customer satisfaction on all projects.* Our projects are completed on time and within budget and meet customer expectations.
- *We act with integrity in all that we do.* We are personally accountable for all that we do, and we act with honesty and fairness.
- *We regard suppliers as essential team members.* We owe our suppliers the same respect that we show to our customers. We treat them fairly and equitably in all business transactions.
- *We regard subcontractors as essential team members.* We owe our subcontractors the same respect that we show to our customers. They are essential team members in the delivery of quality projects. We treat them fairly and equitably.

---

*The names of companies and their employees used in the examples in this book are fictional, and are not associated in any way with any actual companies or persons that may share the same names(s).

> Our mission is to construct high-quality, management-intensive construction projects in the Western States and provide professional, client-oriented services. It is impossible to accomplish this mission without a commitment to the highest standards of ethics and integrity. Our standards of business conduct guide us in making decisions every day. These standards are
>
> - *Relationships*
>   - *With Customers*—We expect our customers to select us because of the quality, the service, and the cost of our services. Our projects are completed to each customer's satisfaction within a mutually agreeable time frame.
>   - *With Suppliers*—We expect our suppliers to provide quality materials within a mutually agreeable time frame. In return, we provide timely payment for all materials received.
>   - *With Subcontractors*—We expect our subcontractors to complete their scopes of work within mutually agreeable time frames. We do not share subcontractor proposals with others but select subcontractors based on the quality of work that they provide. We provide timely payment to all subcontractors, upon receipt of invoices from them.
> - *Conflict of Interest*—We do not engage in any activity that may erode public trust in our company or provide a perception of unfairness in our relationships with others.
> - *Company Resources*
>   - *Time*—Time records are prepared accurately and personal business is not conducted during business hours.
>   - *Resources*—Company resources are not used for personal activities and are safeguarded at all times.
>   - *Information*—Company business practices are not shared outside the company unless required by law. Intellectual property is safeguarded as an essential company resource.

FIGURE 1.4 Statement of Business Conduct for Western Construction

Summary

Construction is an intensely competitive industry with a diversity in the types and in the sizes of companies. Each segment of the industry has its own set of technical challenges, but the basic business mission of a construction company is to obtain the work, to perform the work, and to manage the firm's financial, capital, and human resources.

Construction is a risky business, and about half of the construction firms in the United States fail before they complete ten years in business. The major external causes of business failure are prolonged economic recession, loss of a major customer, new competition, and shortage of skilled employees. The primary internal causes of business failure are pursuit of volume at the expense of profitability, lack of a comprehensive business plan, accepting too large a project, diversifying into unfamiliar types of projects or unfamiliar geographic areas, unplanned loss of key personnel, lack of managerial maturity, poor use of cost-estimating skills, poor use of accounting systems, lack of good equipment control, excessive debt, and poor internal communications. Construction firms need to adopt business strategies to mitigate their risks and minimize the likelihood of business failure.

Construction company chief executive officers and senior leaders set the example for their employees regarding company core values and business ethics. They are responsible for establishing and nurturing the company business culture. Many successful company leaders focus on long-term planning and delegate the day-to-day business operations to subordinates. A major long-term planning issue that senior managers need to focus on is attracting, developing, and maintaining a cohesive, motivated, and skilled workforce.

To be effective, construction firm leaders need to understand how people behave in organizations and how to motivate them to perform well and be committed to the company. Understanding human needs is the key to developing employee motivation strategies. Leaders must understand that employees possess unique personalities and have different attitudes. Personality traits should be considered when selecting an individual for a position within the company. Selecting and developing teams is another essential leadership skill. Little is done in construction by individuals; most tasks are performed by teams.

Review Questions

1. What are the three external influences on business failure within the construction industry?
2. What risks does a construction company assume by taking on an extremely large project; for example, one that requires use of 60 percent of its resources?
3. What strategy do you suggest a construction company adopt when establishing a presence in a new geographical area?
4. What is meant by the term "managerial maturity," and how does it relate to risk of business failure?
5. What strategy do you suggest a utility construction company adopt for the acquisition of a new piece of equipment?
6. What are the four basic management functions that are performed by construction company leaders?
7. What are the two reasons why a construction company president may choose to delegate day-to-day business decisions to subordinates?
8. What are the three primary measures of employee motivation? What are the primary factors influencing each of these measures?
9. What is Maslow's hierarchy of human needs, and how can they be used to develop employee motivation strategies?
10. What are the five basic characteristics of an individual's personality?
11. What is meant by the term "company ethics?"
12. How does a company leader influence the ethical performance of the company's employees?

Exercises

1. You are a senior operations manager for a construction company and are developing a plan to establish a branch office in Utah. Your company currently has no operations in that state. What factors would you consider in developing your plan? Under what conditions would you consider withdrawing from the state?
2. Develop a list of the primary responsibilities for the president of a 75-person construction company.

Sources of Additional Information

Hesselbein, Frances, Marshall Goldsmith, and Richard Beckhard, eds. *The Leader of the Future: New Visions, Strategies, and Practices for the Next Era*, San Francisco, Calif.: Jossey-Bass Publishers, 1996.

Kossoff, Leslie L. *Executive Thinking: The Dream, The Vision, The Mission Achieved*, Palo Alto, Calif.: Consulting Psychologists Press, Inc., 1999.

Schleifer, Thomas C. *Construction Contractors' Survival Guide*, New York: John Wiley & Sons, Inc., 1990.

CHAPTER 2
Company Organization

INTRODUCTION

As was mentioned in Chapter 1, the selection of a form of business organization and of an appropriate organizational structure are basic management functions of construction company leaders. The selection of an appropriate form of business organization is a significant business decision, requiring considerable thought, as each has different tax considerations and liability issues. The most appropriate organizational structure is the one that best fits the business culture, the operating practices, and the business plan created by the company leaders. A company's size, scope of services, and geographic market area also will influence the type of structure selected. In this chapter, we will examine the alternative forms of business organization and the types of organizational structures that may be selected for a company. The organizational structure will determine how employees do their work and establish relationships among organizational elements. Then we will discuss the development of an organization chart and operating procedures for a construction firm.

ALTERNATIVE FORMS OF BUSINESS ORGANIZATION

The basic forms of business organization are shown in Figure 2.1. Each type is found in the construction industry. Selection of the proper type depends on many considerations and is a matter requiring careful study. Each form of organization has its own legal, tax, and financial implications which should be thoroughly examined. Each type of organization has its own advantages and disadvantages. Only federal tax liability is addressed in this chapter. Since each state has its own tax statutes, a construction company owner should consult with a tax adviser to determine state tax liabilities of the alternative forms of business organization.

The selection of an appropriate type of organization for a construction firm is a significant business decision, and company leaders should seek legal and taxation advice when selecting appropriate business organizational structures for their firms. Important factors to consider in selecting a form of business organization are

- The amount of capital needed to establish the company.[1]
- The nature and magnitude of potential liabilities to be faced by the company.
- The number of owners anticipated.
- The extent of the owners' participation in the management of the company.
- The tax consequences of the various forms of business organization.

---

[1]The company's need for capital rather than the number of employees tends to have a greater influence on the selection of an appropriate form of business organization.

Type of Business Organization	Ownership
Proprietorship	Single Owner

Partnership
   General Partnership
   Limited Partnership
	Two or More Owners

Corporation
   Conventional
   Type S
	Multiple Owners

Limited Liability Company	Two or More Owners
Joint Venture	Alliance of Two or More Firms

FIGURE 2.1 Alternative Forms of Business Organization

Proprietorships

Proprietorships are companies that are owned by one individual. They may operate under the name of the owner or may use a trade name. This form of organization is the easiest and least expensive to establish and enjoys the maximum freedom from government regulation. Sole proprietors need not maintain business records except those necessary for tax purposes, because, unlike corporations, they are not required to file annual reports with government agencies. Proprietors must, however, maintain records documenting income and expenses. The organizational structure of a proprietorship is whatever the owner wants it to be. No formal documents are required to establish a business of this type other than registration with appropriate tax authorities, payment of licensing fees, and procurement of required insurance.

The proprietor owns and operates the business, provides the necessary capital, and furnishes all equipment and other capital assets. All business transactions and contracts are executed in the owner's name. The distribution of profits is simple. Any income generated by the company is reported on the owner's personal tax returns and is taxed at the normal individual tax rates. The owner is personally liable for any liabilities incurred by the company. This unlimited liability extends to the owner's personal assets, even though they may not be involved in the business. The individual and the proprietorship are not recognized as being separate under the law, even if the business operates under a trade name.

The proprietorship continues until sold, abandoned, or the death of the proprietor. Continuity of operation in the event of the death of the proprietor may be achieved by a direction in the proprietor's will; that is, if the executor is directed to operate the business until sold or transferred to the individual specified in the will.

The major advantages and disadvantages of proprietorships are

Advantages

- Easy and inexpensive to establish
- Minimal start-up costs
- Total control by owner

Disadvantages

- Unlimited liability of owner
- Business terminates on death of owner
- Difficult to attract capital due to lack of successor management

General Partnerships

State laws regarding partnerships are not the same, although most are based on the Uniform Partnership Act.[2] A **partnership** is an association of two or more persons to conduct business as co-owners. Although it has its own assets and conducts business, a partnership, unlike a corporation, is not treated as a legal entity separate from the individuals forming the partnership. However, in some jurisdictions, a partnership may own land, hire employees, and sue or be sued. This pooling of financial resources offers the potential for conducting larger construction operations than would be possible for the partners individually.

A partnership is not recognized by law as being an entity separate from the partners. Each partner contributes capital or other assets to the partnership. The partners generally share management responsibilities and share the profits of the partnership in proportion to the relative value of their contributions to the firm. A partnership can be formed based on an oral agreement, but it is highly desirable to have a written agreement that states the rights, the responsibilities, and the obligations of each partner. Legal advice should be sought when creating the partnership agreement.

A partnership pays no taxes, although it must file an information tax return. Profits are taxed at the individual tax rate of each partner. The partners have joint control of the company. They typically are paid salaries and annually receive proportionate shares of the profits of the partnership. Each partner is an agent for the partnership and is able to enter into binding contracts in the name of the partnership, unless otherwise stipulated in the partnership agreement.

Each partner assumes unlimited personal liability to third parties for any contracts and debts of the partnership, irrespective of the value of the partner's contribution to the partnership. The partners are jointly and severally liable for all partnership obligations and liabilities, including any misrepresentation perpetrated by another partner in the conduct of the business affairs of the partnership. Each partner accepts unlimited financial responsibility for the acts of the other partners and underwrites the liabilities of the partnership to the full extent of his or her personal financial resources. This means that if any one partner is unable to pay his or her share of the partnership's liabilities, creditors can force the remaining partners to pay the share of the first. Consequently, one should exercise great care in the selection of partners when organizing a partnership.

A partnership continues until the death or withdrawal of one of the partners. If each partner has executed a buy/sell agreement with the other partners, the partnership may continue following the death or withdrawal of a partner. These

---

[2]The Uniform Partnership Act was approved by the National Conference of Commissioners on Uniform Laws in 1914 and revised in 1994.

buy/sell agreements specify how the value of a partner's share of the business is to be determined and the right of the remaining partners to purchase the departing member's share. Life insurance policies are often used to provide the financial resources needed to purchase a deceased partner's interest according to the buy/sell agreement. In such circumstances, the partnership would be the named beneficiary on the insurance policy. The partnership may also be dissolved by written agreement among the partners, and individual partners may sell their interests in the partnership to the remaining partners.

The major advantages and disadvantages of general partnerships are

Advantages

- Pooling of financial resources
- Pooling of talent
- Sharing of management responsibilities
- Easier to attract capital than proprietorships

Disadvantages

- Unlimited liability of partners
- Any partner can obligate partnership
- Business may terminate on departure or death of partner

Because of the unlimited liability of the partners, a new form of partnership, known as a limited liability partnership, has been created. Such partnerships limit the partners' liability to the assets of the partnership.

Limited Partnerships

Under the Uniform Limited Partnership Act,[3] a **limited partnership** may be formed with two categories of partners, general partners and limited partners. A general partner contributes resources to the partnership and performs a management function, while a limited partner also contributes resources to the partnership, but has no voice in the management of the partnership. The operation of a limited partnership, as well as its establishment, must be in accordance with the laws of the state in which it is created.

The primary purpose for forming a limited partnership is to raise capital without forming a corporation. Limited partners are allowed to invest in the partnership, to receive profits from the business, and to limit their liability to the amount of their investments. Profits are distributed to all partners and taxed at the individual tax rate of each partner. The liability of the limited partners is limited to their investments in the partnership. The general partners, on the other hand, bear unlimited liability for any obligations of the partnership.

As with general partnerships, a limited partnership may terminate on the departure or the death of a general partner, unless buy/sell continuity agreements have been executed. The limited partnership, however, is not automatically terminated upon the departure or the death of a limited partner.

---

[3]The Uniform Limited Partnership Act was approved by the National Conference of Commissioners on Uniform Laws in 1916 and revised in 1976.

The major advantages and disadvantages of limited partnerships are

Advantages

- Pooling of financial resources
- Pooling of talent
- Sharing of management responsibilities
- Limited liability of the limited partners
- Easier to attract capital than general partnerships or proprietorships, because of the limited liability of limited partners

Disadvantages

- Unlimited liability of general partners
- Any general partner can obligate partnership
- Business may terminate on departure or death of a general partner

Corporations

A **corporation** is a legal entity formed under state law by a certain number of stockholders filing a certificate of articles of incorporation with an appropriate official of the state government, and it operates under a corporate name. The corporation also develops bylaws to govern its day-to-day business operations. The state then grants a charter to the corporation allowing it to operate under the terms of its charter (articles of incorporation), its bylaws, and applicable state laws. The formation of a corporation brings into being a legal entity separate and distinct from the owners of the corporation. It owns all assets and owes all debts of the company.

There are two basic types of corporations, privately held and publicly held. Some privately held corporations are family-owned, while others are employee-owned. An example of a family-owned corporation is Bechtel, and an example of an employee-owned corporation is Kiewit. Publicly owned corporations usually are listed on a stock exchange and owned by individuals who may or may not be associated with the company. The individual stockholders generally have immunity from all corporate liability, which makes incorporation an attractive form of business enterprise.

Corporations enjoy the right of perpetual succession and are regarded by law as being separate and distinct from the owners. Corporations are authorized to conduct business, to own property, to employ people, to enter into contracts, and to incur debt. The owners of a corporation are known as stockholders, in that they own stock in proportion to their investments in the corporation. The profits of a corporation are subject to income taxes at the corporate rate. Such taxes are paid before any profits are distributed to the stockholders in the form of dividends. The stockholders pay taxes on any dividends provided by the corporation, which results in double taxation of those corporate profits that are distributed to the stockholders.

The stockholders elect a board of directors to exercise general control over the corporation. The board of directors appoints the officers of the corporation, who exercise day-to-day management of the corporation, following the procedures described in the corporation's bylaws. Senior managers may or may not be elected members of the board of directors.

A major advantage of a corporation is that it allows individuals to pool their resources on a limited-risk basis in a profit-seeking business that is managed by people of their choice. The maximum liability of the stockholders is the value of their investment in the corporation. The owners of a corporation may or may not be involved in the management of the organization. A primary stockholder may be involved in managing the corporation, or the corporation may be managed by individuals who own little or no stock. Another advantage of a corporation is that in most states it has a perpetual term of existence beyond the death of a stockholder. Ownership in the corporation is easily transferred by the sale of shares of stock.

A corporation is the most expensive type of organization to create and is subject to more government regulation than are partnerships or proprietorships. Extensive record keeping is required, which is subject to government review.

The major advantages and disadvantages of corporations are

Advantages

- Limited liability of owners
- Pooling of talent
- Sharing of management responsibilities
- Perpetual life of company
- Easier to attract capital than partnerships or proprietorships, because investors need not be employed by the corporation

Disadvantages

- Closely regulated and state charter required
- Most expensive business form to create
- Extensive record keeping required
- Double taxation of profits distributed as dividends

Type S Corporations

Closely held corporations that meet Internal Revenue Service requirements may choose to be treated as a **type S corporation**. Such corporations are taxed as partnerships rather than as corporations. The number of stockholders allowed is limited to 75, there must be only one class of stock, and all stockholders must consent to the company being treated as a type S corporation. Stockholders enjoy the corporate benefits of limited liability and the ability to transfer ownership, but there is no double taxation on dividends. Corporate profits are taxed at the individual rate of each stockholder instead of at the corporate rate. A start-up company could go in for a type S corporation, because company losses can be used to offset shareholders' other taxable income.

The major advantages and disadvantages of type S corporations are

Advantages

- Limited liability of owners
- Pooling of talent
- Sharing of management responsibilities
- Perpetual life of company
- Easier to attract capital than partnerships or proprietorships
- No double taxation of corporate profits

Disadvantages

- Closely regulated and state charter required
- Most expensive business form to create
- Extensive record keeping required
- Limitation on number of stockholders and business volume

Limited Liability Companies

A **limited liability company** is a legal entity, similar to a corporation, which combines limited personal liability with pass-through tax benefits, similar to the type S corporation. Such companies are formed by two or more people filing written agreements with an appropriate official of the state government. Such agreements must be prepared in accordance with applicable state statutes. A limited liability company is different from a limited partnership in that there are no general partners. Also, it is different from a corporation because it has no stockholders or corporate officers. Income generated by a limited liability company is taxed in a manner similar to a partnership. The states impose less regulatory requirements on limited liability companies than they do on corporations. In contrast to a corporation, the limited liability company can be dissolved upon the death or retirement of an owner. Some states allow the creation of limited liability partnerships for certain professions, such as doctors or lawyers, but not construction companies.

The major advantages and disadvantages of limited liability companies are

Advantages

- Limited liability of owners
- Pooling of talent
- Sharing of management responsibilities
- Easier to attract capital than partnerships or proprietorships
- Less regulated than a corporation

Disadvantages

- Regulated and state charter required
- More expensive business form to create than a partnership or proprietorship

Joint Ventures

A **joint venture** is a business alliance between two or more business enterprises. The members of a joint venture may be proprietorships, partnerships, or corporations. Often, joint ventures are companies that group together temporarily to perform a specific project. Joint ventures are formed to share risk, to acquire expertise, or to provide adequate resources to construct the project. They combine the resources, assets, and skills of the participating companies. Each member participates in the management of the joint venture, shares the risk, and receives a portion of the profits generated.

An example would be a joint venture between a design firm and a construction firm to perform a design-build project—such as Wasatch Constructors, formed by

Kiewit Pacific, Granite Construction, and Washington Construction, to design and construct a major section of Interstate 15 in Utah prior to the 2002 Winter Olympics. Some joint ventures are long-term relationships that provide a melding of expertise to compete in a particular market. A joint venture can be a partnership or a corporation, but the most common is a special-purpose partnership that is organized to undertake a single project. These special-purpose partnerships typically are terminated upon the completion of the project. Taxing and general liability depend on the type of structure chosen for the joint venture.

The participants in the joint venture enter into a written agreement that defines the objectives of the organization and describes the obligations and the rights of each party, the percentage interest of each participant, the amount of working capital to be contributed by each party, the responsibilities and the procedures for managing the joint venture, and the limitations on liability. There is no limitation on the liability between the joint venture and the participants. If there is a default by any member of the joint venture, the remaining members are responsible for any contractual requirements. Crafting a good joint venture agreement requires skilled legal advice.

The major advantages and disadvantages of joint ventures are

Advantages

- Pooling of financial and equipment resources
- Pooling of talent and expertise
- Sharing of management responsibilities
- Sharing of risk

Disadvantages

- Loss of control
- Sharing of profits

ALTERNATIVE ORGANIZATIONAL STRUCTURES

Once the type of business organization has been selected for a construction firm, the next decision to be made is the organizational structure for the company. Before selecting an organizational structure, company leaders need to decide the following:

- how the firm is to accomplish its mission,
- the work processes to be used by employees, and
- the authority relationships to be used to manage the company.

The challenge is to create a suitable organizational structure that provides an adequate field structure to manage the company's projects and a company administrative structure that serves in a support capacity for all company operations. The more common types of organizational structures are functional, divisional, and matrix. Each has its advantages and disadvantages, which should be considered in making a selection.

Functional Organization Structure

Now let's examine the three alternative organizational structures. The first is a **functional organization**, which is illustrated in Figure 2.2. The functional structure is the most common type of organizational structure used by most small and mid-size[4] construction firms. It uses organizational elements that perform specialized functional activities such as construction operations or business development. The basic characteristics of a functional organization are

- Organizational breakdown based on function or process.
- Line and staff functions—Line functions are those jobs that directly affect the principal work flow, and staff functions are those jobs that provide service or advice to line organizational elements.
- Chain of supervision—Authority and responsibility are arranged hierarchically with only one supervisor for each position.

The major advantages and disadvantages of the functional organization are

Advantages

- Promotes skill specialization
- Uses human resources efficiently
- Enhances specialized career development
- Facilitates vertical communication and centralized decision making

Disadvantages

- Fosters parochial perspectives
- Inhibits cross-functional communication
- Obscures accountability
- Enhances potential for functional conflicts

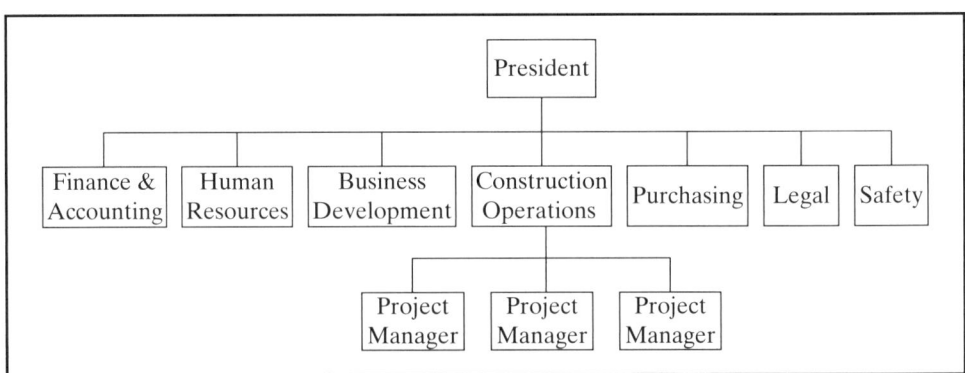

FIGURE 2.2 Sample of Functional Organization

---

[4]A construction company with less than 50 employees would be considered small, and one with 50 to 200 employees would be considered mid-size.

The functional organization structure provides clear assignment and identification of responsibilities but creates a need for establishing integrating mechanisms to coordinate the activities of the functional elements. It encourages limited perspectives among employees focused on a narrow set of functional tasks rather than on the overall company strategic goals. Such an organizational structure may be difficult to use if the company serves multiple geographic areas or offers a variety of services. A primary advantage is its efficient utilization of people.

Divisional Organizational Structure

The second type of organizational structure is a **divisional organization**, which is illustrated in Figure 2.3. In this type of organization, there may be multiple functional elements, such

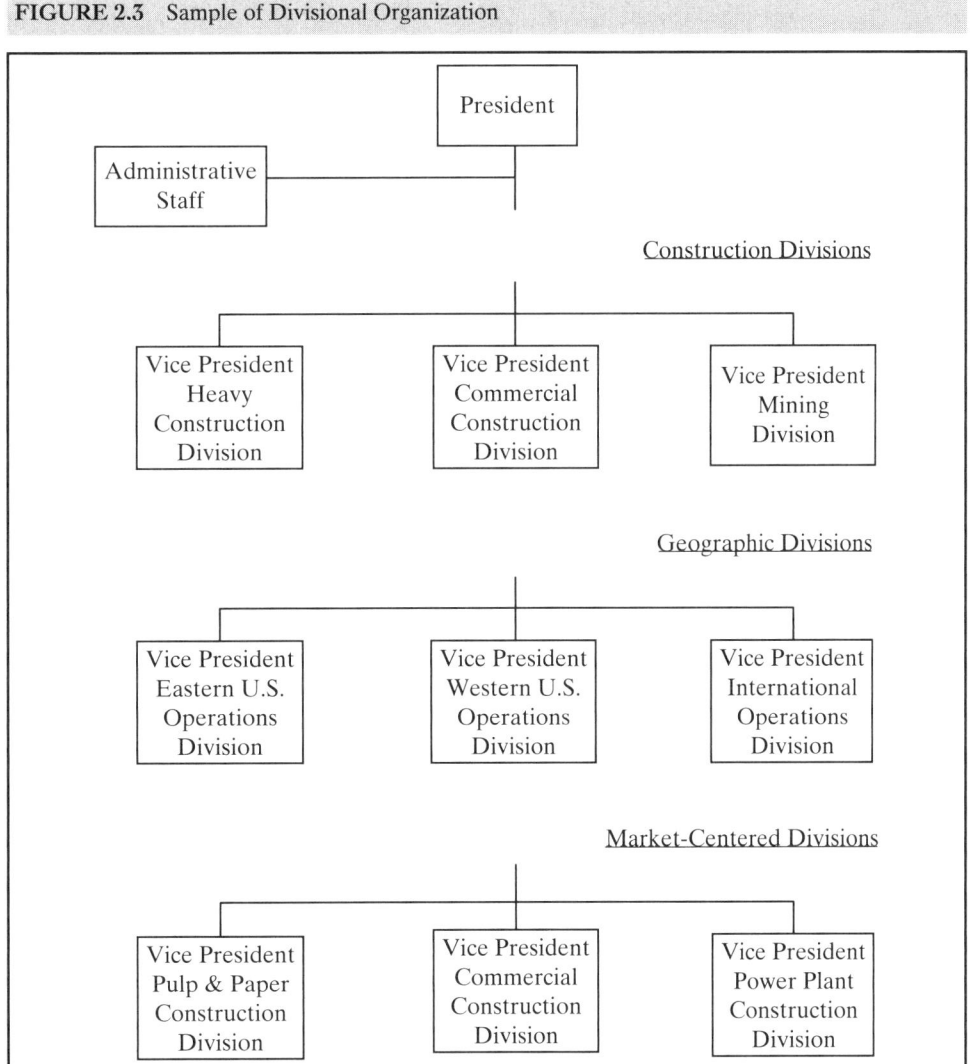

FIGURE 2.3 Sample of Divisional Organization

as a construction operations section in each division. Divisions may be based on geographical regions, types of customers, or categories of projects. This type of organization would be used primarily by large construction firms who operate in many locations or work across numerous segments of the market. Divisional organizations often start as functional organizations but evolve as the number of markets or services increases. Most national and international construction firms use some type of a divisional structure.

The major advantages and disadvantages of the divisional organization are

Advantages

- Focuses on products, customers, and markets
- Recognizes interdepartmental interdependencies
- Fosters cohesion
- Ensures accountability
- Allows diversification
- Increases strategic and operational control
- Minimizes problems due to sharing of resources across functional areas

Disadvantages

- Uses resources inefficiently
- Limits career advancement within functional areas
- May lead to dysfunctional competition among divisions

The divisional organizational structure reduces the integration problem existing in the functional structure by focusing self-contained organizational elements on specific markets. A major leadership challenge with divisional organizations is to ensure that information is shared among the various divisions.

Matrix Organizational Structure

The third type of organization is a **matrix organization**, which is illustrated in Figure 2.4. This type of structure is used in decentralized organizations that assign employees to multifunctional teams while retaining functional chiefs. This is a combination of the functional and departmental structures and is the most difficult organizational structure to manage. Managers of functional and service departments share the supervision of subordinates. The matrix structure typically evolves. The initial stage is the creation of temporary task forces to handle particular issues or customers. The second stage is the creation of permanent teams. The final stage is when project managers are appointed and held accountable for integrating the teams' activities. Matrix organizations are not as common in construction as are functional and divisional organizations. As illustrated in Figure 2.4, a matrix organization may be created to provide technical oversight of individuals assigned to various divisions.

The advantages and disadvantages of a matrix organization are

Advantages

- Makes specialized knowledge available to all projects
- Uses resources flexibly to meet project requirements
- Forces communication among managers
- Can be adapted by shifting emphasis between functional and project orientation

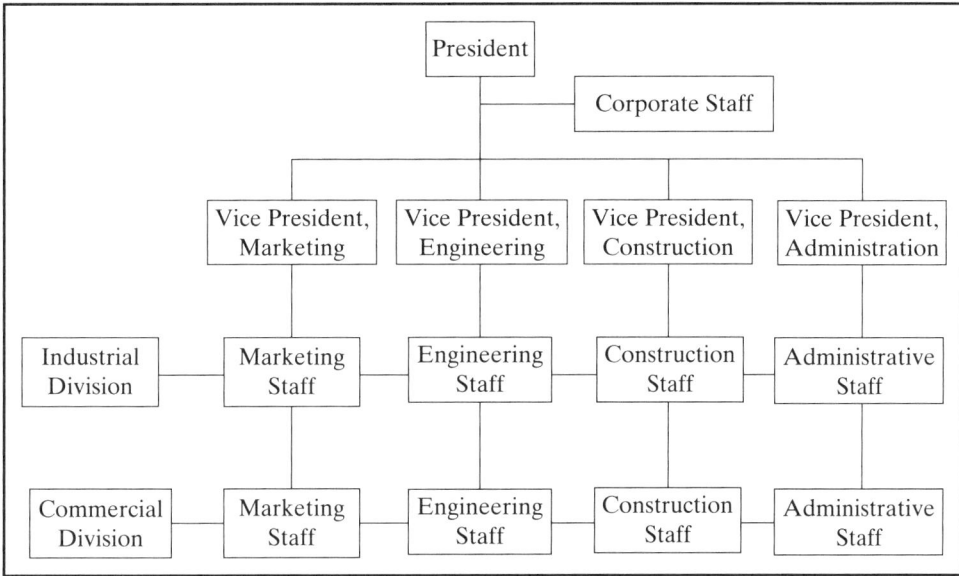

FIGURE 2.4 Sample of Matrix Organization

Disadvantages

- Increases stress due to dual reporting relationships
- May lead to power struggles between functional and project managers

Matrix organizations generally require the committed managers and subordinates to be effective, because of dual supervisors for many team members.

ORGANIZATIONAL DESIGN

Organizational design is the process of selecting an organizational structure for the company and a formal system of communication, authority, and responsibility necessary to achieve the company's goals and objectives. A good design should

- Facilitate the flow of information and decision making to manage uncertainty and achieve company goals.
- Define the authority and responsibility assigned to organizational elements so that the benefits from the division of labor and effective job design can be realized.
- Create the desired degree of integration among organizational elements.

Major considerations in selecting an appropriate organizational design are

- Work flow
- Division of labor
- Task interdependence

- Hierarchy of authority
- Business procedures and processes
- Interdepartmental relations

Work flow involves analyzing the way in which the work is performed within the company to accomplish its strategic goals. *Division of labor* considers the extent to which jobs will be specialized within the company. *Task interdependence* involves analyzing the relationships among the various tasks that are performed within the company. *Hierarchy of authority* involves determining the amount of position and task grouping that will be used and the levels where business decisions will be made. *Business procedures and processes* involve determining how business operations will be conducted, including the use of information management technology. *Interdepartmental relations* involve assessing how the various organizational elements should relate to each other.

Through the design of an organizational structure, company leaders decide how the company's mission will be accomplished and how employees and functions will be grouped. The resulting structure will prescribe task and authority relationships within the firm as well as a formal system of communication. The first set of decisions relate to the type of organizational structure to be used by the company.

- Is the company to use a functional, divisional, or matrix organization?
- Is it to be a pyramid or a flat organization, which relates to the span of control of leaders within the company? **Span of control** means the number of people supervised by each manager.
- Are decisions made centrally, or is decision making decentralized? Companies with centralized decision-making processes tend to adopt smaller spans of control for company managers than do companies that adopt decentralized decision-making processes. A centralized company might select spans of control of 5 to 8 people, while a decentralized company may adopt spans of control of 10 to 15 people.

Other factors that should be considered are

- Geographical area served by the firm
- Anticipated number of employees
- Type of products and services to be offered
- Type of customers to be served

Once the basic approach to developing the organizational structure has been selected, the next step is to develop an organizational chart for the company. This involves analyzing the way in which work is performed within the firm and answering the following questions:

- Which tasks are performed at field offices, at regional offices, and at the home office? A listing of home office and field office functions for a typical construction company is shown in Figure 2.5. Field office functions typically are those needed for direct management of construction projects, while those performed at a home office are those needed for management of the company. Generally the number of functions performed in field offices is kept to a minimum to reduce project overhead costs.

CHAPTER 2　Company Organization　29

Home Office Functions	Field Office Functions
Accounting and Financial Management	Contract Administration
Business Development	Equipment Control
Central Procurement	Project Management
Equipment Management	Project Procurement
Safety Program Management	Safety Management
Cost Estimating/Bidding	Project Cost Accounting
Human Resources Management	Quality Control Management

FIGURE 2.5 Sample Allocation of Construction Firm Functions

- What is the anticipated workload in each functional area of each organizational element?
- How many people are required in each organizational element to accomplish the anticipated workload efficiently?

An **organization chart** is a pictorial representation of the company's organizational structure that shows functional relationships as well as lines of supervision and authority. Most construction companies have an operations activity, which may include field offices, to manage the actual construction of projects. All business functional activities typically are grouped in a central office with needed project support services. The organization structure should be designed to accommodate growth, but be flexible and adaptable, as demand for construction services is often cyclical. The organization of the central office needs to be efficient and economical, to minimize the overhead cost of the company. In small firms, individuals may have multiple responsibilities, while in larger firms with more business volume, there may be multiple people performing similar responsibilities.

An organization chart for a small proprietorship may look like the one illustrated in Figure 2.6. An organization chart for a partnership may look like the one illustrated

FIGURE 2.6 Organization Chart for Small Proprietorship

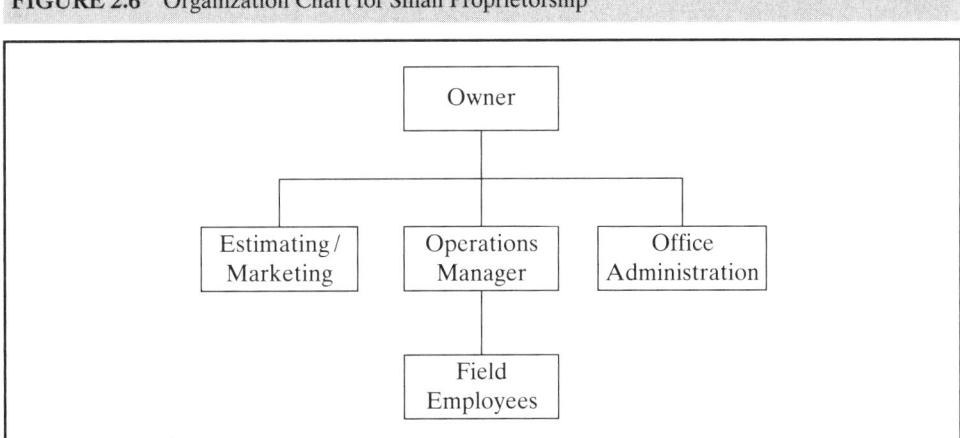

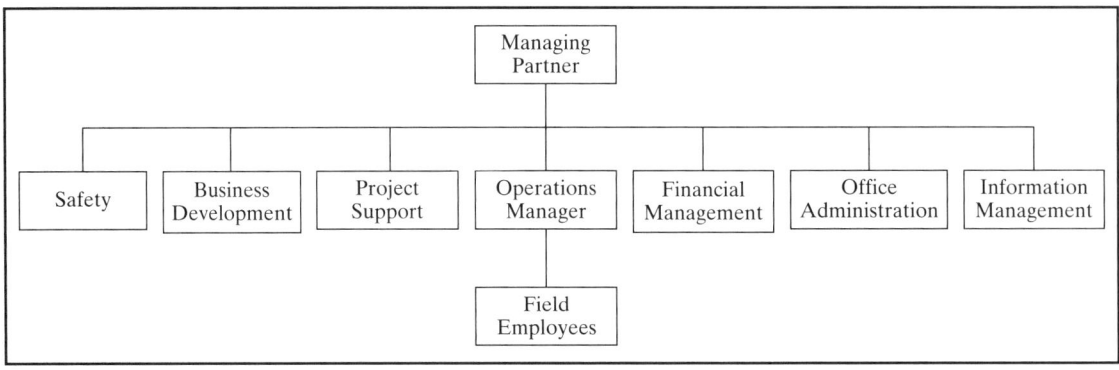

FIGURE 2.7 Organization Chart for Partnership

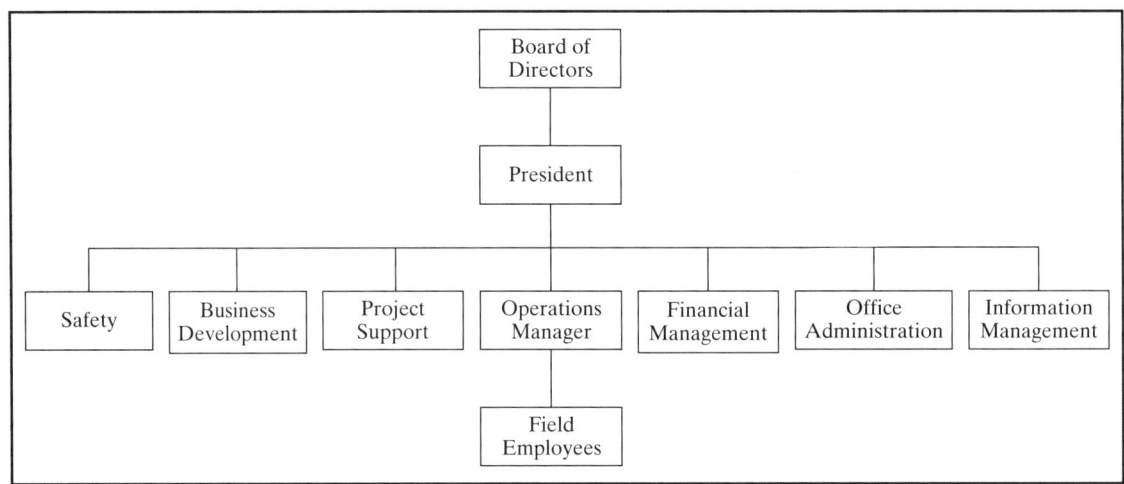

FIGURE 2.8 Organization Chart for Corporation

in Figure 2.7. An organization chart for a corporation may look like the one illustrated in Figure 2.8.

Once the organization chart has been developed, the number of people needed for each function needs to be determined and job descriptions prepared for each position. We will discuss both of these topics in Chapter 7 as a part of our discussion of human resources management.

POLICIES AND OPERATING PROCEDURES

Once the organization structure has been selected, the next issue to be addressed is development of written company policies and operating procedures. These policies and procedures provide guidelines for all levels of company operations and provide guidance for decision making in recurring situations. Operating procedures establish general rules governing communications (written and electronic), the flow of paper, and other routine company operations. The primary purpose of the document is to standardize operating procedures and decentralize decision making in routine events to allow company leaders to focus on more critical issues.

The manual is often divided into sections, with each section addressing the policies and procedures related to the individual functions listed in Figure 2.5. An important section is control of company overhead costs, including annual budgeting and monthly management. Such manuals usually also describe the composition and frequency of various meetings convened to address company business management issues.

The policies and procedures manual should also contain forms or electronic formats to be used, assign responsibilities to individual organizational elements, and assign signature authority for various documents. An example of a signature authority policy is shown in Figure 2.9. Such policies are established to indicate authority for financially obligating the company. Documents that have significant financial impact on the company are usually reserved for approval and signature by a corporate officer as part of the company's risk management policy.

The manual should also describe all records that are to be retained and how long they should be retained as well as all reports that are to be prepared, the frequency for each report, and to whom the reports are to be submitted. Company leaders need timely reporting of key company performance data to support their decision-making processes.

FIGURE 2.9 Sample of Signature Authority Policy

CONTRACTS, SUBCONTRACTS, AND CONTRACT MODIFICATIONS	
Customer/Company	President
Company/Subcontractors	Corporate Officer
Change Orders	Corporate Officer
Purchase Orders over $100,000	Corporate Officer
PROJECT MANAGEMENT, ENGINEERING, AND ACCOUNTING	
Schedule of Values Submission	Project Manager
Payment Applications	Corporate Officer
Subcontractor Schedule of Values Approval	Project Manager
Shop Drawing Review/Transmittals	Project Engineer
Schedule Transmittal/Notifications	Project Manager
Lien Releases	Project Manager
Purchase Orders under $100,000	Project Manager
DOCUMENTATION AND CORRESPONDENCE	
Job Meeting Minutes	Project Manager
Routine Correspondence	Project Manager
Delay Notice	Project Manager
Request for Change Order	Project Manager
Change Order Proposal	Project Manager
Notification of Work under Protest	Project Manager
Work Stoppage	Corporate Officer
Notification of Claim	Corporate Officer
FINANCIAL INSTRUMENTS	
Checks	Corporate Officer
Loan Applications	Corporate Officer

Summary

The selection of a form of business organization and an appropriate company organizational structure are basic management functions of the company leadership. Selection of an appropriate form of business organization is a significant decision, influenced by liability and tax considerations. The most appropriate structure is the one that best fits the operating culture and business practices of the firm.

A proprietorship is a company owned by one individual. All profits are taxed at the owner's personal tax rate, and the owner has unlimited liability for the firm. A partnership is an association of two or more persons to conduct business as co-owners. In a general partnership, all partners share the tax liability for profits and the unlimited liability for the firm. In a limited partnership, all partners share the tax liability for profits, but only the general partners have unlimited liability for the firm. The limited partners are liable only up to the value of their investments in the firm.

A corporation is a separate legal entity formed for the purpose of conducting business. Profits are taxed at the corporate rate, and owners (stockholders) are liable only for the value of their investments in the firm. A joint venture is a business alliance between two or more firms. It may be a temporary alliance for the purpose of completing a specific project or a long-term relationship to compete in specific markets.

Once the type of business organization has been selected, the next decision is to select an organizational structure for the company. The alternatives are functional, divisional, or matrix. Each has its own advantages and disadvantages, which should be carefully considered in selecting the structure for the firm. Once the organizational structure has been selected, specific operating procedures need to be developed for the company.

Review Questions

1. What is the difference between a proprietorship and a partnership?
2. What is the difference between a general partnership and a limited partnership?
3. What is the difference between a partnership and a corporation?
4. What is the difference between a type S corporation and a limited liability company?
5. What is a joint venture and why might two construction firms choose to form one?
6. What is a functional organization and what are its advantages and disadvantages?
7. What is a divisional organization and what are its advantages and disadvantages?
8. What is a matrix organization and what are its advantages and disadvantages?
9. What are the major factors to be considered in designing an organizational structure for a construction company?
10. What issues should be addressed in a construction company's operating procedures manual?

Exercises

1. Develop an organization chart for a mid-size local construction company that provides both construction and construction management services to clients. The number of employees is 80, and the annual volume of work is $50 million per year.
2. Prepare a table of contents for an operating procedures manual for the company discussed in Question 1.

Sources of Additional Information

Bockrath, Joseph T. *Contracts and the Legal Environment for Engineers & Architects*, 6th ed., New York: McGraw-Hill, 2000.

Clough, Richard H., Glenn A. Sears, and S. Keoki Sears. *Construction Contracting*, 7th ed., Hoboken, N.J.: John Wiley & Sons, Inc., 2005.

Hellriegel, Don, John W. Slocum, Jr., and Richard W. Woodman. *Organizational Behavior*, 5th ed., St. Paul Minn.: West Publishing Company, 1989.

Ivancevich, John M. and Michael T. Matteson. *Organizational Behavior and Management*, 4th ed., Chicago: Irwin, 1996.

Sweet, Justin and Marc M. Schneier. *Legal Aspects of Architecture, Engineering, and the Construction Process*, 7th ed., Toronto: Thompson, 2004.

CHAPTER 3

Risk Management

INTRODUCTION

As discussed in Chapter 1, construction is a risky business, and risk management is an essential responsibility in managing a construction company. This includes the risks encountered by being a business enterprise as well as the risks associated with constructing projects. Business risks include liability for the activities of employees and potential loss of or damage to company property. When a construction firm executes a construction contract with a project owner, it assumes the risks associated with constructing the project, as defined in the project plans and specifications.

Depending on how the contract is priced (lump sum, unit price, or cost plus), the construction company may have assumed the risk of obtaining the needed materials, labor, and equipment within an agreed price and completing the project by the contractual completion date. Some of this risk may be transferred contractually to subcontractors, but much of the risk remains with the general contractor. In addition, the construction company incurs the risk of subcontractor bankruptcy and poor workmanship when using subcontractors.

The major business risks faced by most construction companies are

- health and safety of its employees while working on the job or at another location,
- injury or property loss to third parties as a result of company activities,
- damage or loss to construction work in place but not yet accepted by the owner, and
- loss of or damage to its vehicles and construction equipment.

These risks and liabilities are so great that most construction firms must purchase insurance policies to protect, or partially protect, themselves against financial loss that may result from these risks. In the first part of this chapter, we will discuss the various forms of insurance that are purchased by most construction firms.

In the second part of the chapter, we will discuss bonding. Surety bonds are used as risk management tools by many project owners, and construction companies must be able to provide the required bonds, if the companies are to qualify to compete for such projects. General contractors may also require such bonds from subcontractors to mitigate the risks assumed by subcontracting elements of the work. Surety bonds are not a type of insurance, but are guarantees of performance that are an important aspect of the construction business that must be understood by construction company leaders.

CHAPTER 3   Risk Management   35

RISK MANAGEMENT STRATEGIES

Because construction is a risky business, risk management is a major responsibility of the leaders of a construction firm. The major risk categories are

- People
- Property
- Process

People include both people working on a construction site as well as those who might be impacted by construction operations. Property includes the project under construction, the construction company's property, and property owned by others that might be affected by construction operations. Process means the business procedures used by the construction firm, such as proper disposal of contaminated material.

Risk management involves

- identifying each risk,
- assessing the potential impact on company assets, and
- selecting an appropriate management strategy, as illustrated in Figure 3.1.

Once strategies have been selected, it is important to monitor the results obtained. Some construction firms have a full-time risk manager, while in other companies, risk management is a part-time responsibility, often assigned to the company safety

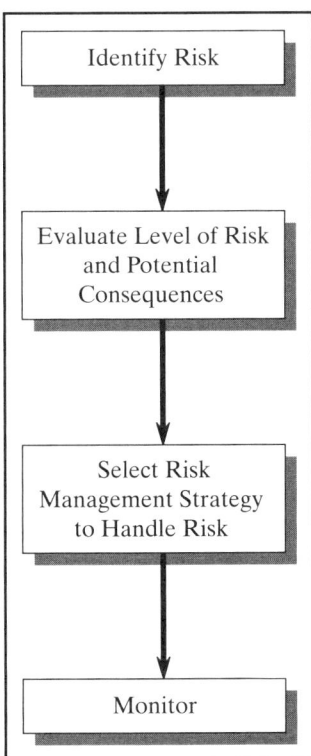

FIGURE 3.1   Risk Management

director. The overall objective of risk management is to reduce the cost of doing business.

In general, there are three alternative strategies for managing a risk:

- Avoid the risk
- Accept the risk
- Transfer the risk

Sometimes a project is too risky to undertake, and a construction firm may choose not to compete for it. This is an example of avoiding the risk. In another project, the construction firm may choose to accept the risk and include an appropriate contingency in its proposal. Risk transfer occurs through contracts, either by language in the construction contract or subcontracts, or by insurance contracts purchased from insurance companies. Requiring surety bonds from subcontractors is another form of risk transfer.

We will discuss the typical types of insurance contracts in the next section. By purchasing insurance, a construction company can

- Transfer risk to a professional risk taker (insurance company).
- Combine risk transfer with risk assumption by the use of deductibles.
- Protect against severe financial loss associated with risk.
- Spread the cost of risk over time.

Accidents on project sites are a major risk to construction firms. Most construction companies minimize their risks by adopting and enforcing good safety programs, but the risk of an accident remains. While the liability for many of the direct costs associated with accidents may be transferred by the purchase of appropriate insurance policies, most of the liability for the indirect costs is not transferable. In many accidents, the indirect cost may be 10 to 20 times the direct costs. The direct costs include medical care for injuries and the cost for repairing property damage. The indirect costs include the loss of production on the project, the adverse publicity on the reputation of the construction company, and third-party lawsuits.

INSURANCE

Insurance Contracts

As mentioned in the last section, a construction company may wish to transfer some of its risk to an insurance company or may be required to do so by the terms of individual construction contracts or by statutes. This involves the purchase of insurance policies. The primary parties to an insurance policy (contract) are the purchaser of the contract (the *insured*) and an insurance company that provides the required insurance coverage (the *insurer* or *carrier*). Additional parties are sometimes named as **additional named insureds** and provided coverage by the policy.

Insurance policies are two-party contracts under which the insurer promises, for a fee or premium, to assume financial responsibility for specified losses or liabilities of the insured for a specified period of time. The amount of risk transferred is limited to the amounts stipulated in the individual insurance policies. Some policies are written with deductible amounts that are the responsibility of the insured. In such policies, the insurance company is liable only for the losses incurred that exceed the deductible amount. For example, if the policy includes a deductible amount of $100,000, the insurance company is liable only for the increment of the loss that exceeds $100,000. The premiums for many types of insurance are adjusted up or down according to the construction company's loss experience record. Premiums for policies that contain deductible amounts are typically less than that for comparable policies with full coverage.

One of the most important areas with respect to insurance policies is to be absolutely certain that the named insured is correctly shown. If an individual company is doing business, it should be so named on the policy. Corporate entities must be named. Joint ventures or partnerships must be named. Just because the individual members of a joint venture have insurance, it does not mean that the joint venture is insured unless it is specifically identified in the policies.

Construction companies should work with their insurance agents to select financially strong insurance companies with demonstrated interest in providing construction insurance. Insurance companies are rated by the A.M. Best Company, using the ratings shown in Figure 3.2. Most construction companies select insurance companies that are rated A or better, but in some circumstances insurance companies rated as B++ or B+ may be used. Using lower-rated companies is not recommended.

Insurance Agents

Construction insurance is a highly specialized and complex field. There is no all-encompassing policy or contract that will cover all insurable risks. Instead, construction insurance is a web of policies that are stitched together to cover the relatively limited number of risks that are insurable. Because construction insurance is specialized and complex, construction companies rely on insurance brokers or agents who are familiar with the risks faced in the construction industry and understand the types of insurance coverage needed by a construction company. These **insurance agents** help the company leaders to develop a comprehensive insurance program and to procure the needed policies at competitive prices. Insurance cost may account for up

FIGURE 3.2 Insurance Company Ratings

Insurance Company Rating Scale used by A. M. Best Company	
Ratings	**Categories**
Secure Ratings:	
A++ and A+	Superior
A and A−	Excellent
B++ and B+	Very Good
Vulnerable Ratings:	
B and B−	Fair
C++ and C+	Marginal
C and C−	Weak
D	Poor

to 20 percent of a construction company's total overhead cost, so it must be carefully managed.

An enduring relationship with a good insurance agent is essential for success in the construction industry. There may be a good reason for changing agents, but rarely do construction firms do so. They may change carriers for individual policies, but typically maintain a long-term relationship with a single agent. A construction firm needs an agent who understands the company's business plan and its insurance needs.

When selecting an insurance agent, a construction company should select an individual or a firm that has extensive knowledge of the construction industry and the risks and challenges faced by construction companies. The job of the insurance agent is to be a source of information regarding

- the types of insurance products on the market,
- the suitability of specific insurance companies,
- the anticipated cost of various levels of coverage, and
- the use of deductibles.

Good insurance agents will provide the following services:

- Program administration
  - Monitor the insurance industry and keep the construction company appraised of changes or trends.
  - Assess construction insurers with respect to service and cost.
  - Review insurance policy language to ensure that the construction firm has the desired coverage.
- Risk management
  - Assist the construction firm in the development of a risk management plan.
  - Respond to coverage questions and any new risk exposure.
- Claim support
  - Act as liaison between the construction firm and the insurance company on any claims issues.
  - Review coverage issues and, if necessary, negotiate with the insurance company on behalf of the construction firm.

Insurance Types

Let's now discuss the major types of construction insurance. Typical insurance policies can be grouped into four areas:

- Property insurance for the construction project
- Property insurance for the construction company's property
- Liability insurance
- Business insurance

Individual policies and coverage are discussed in the remainder of this section.

Property Insurance for the Construction Project

Multiple insurance policies may be needed to provide protection against financial loss due to damage to a project under construction, as illustrated in Figure 3.3.

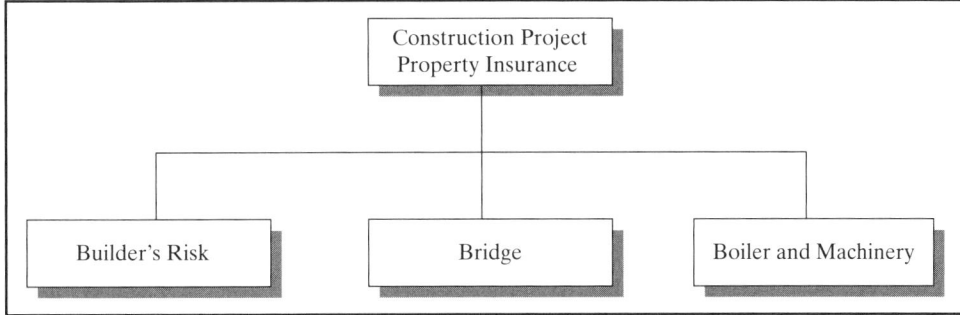

FIGURE 3.3 Insurance Coverage for Construction Projects

Construction company leaders need to carefully review the language of each policy with their insurance agents to ensure that adequate coverage is provided.

Builder's Risk Insurance Builder's risk insurance is a first-party insurance that provides coverage for the cost of damages to the project during construction, which include temporary structures and any materials stored on the project site, but not the contractor's equipment. It protects both the owner and the contractor against direct losses. If listed as named insureds, the insurance also protects all subcontractors working on the project.

Builder's risk insurance is often purchased by the owner, with coverage extending to both the general contractor and the subcontractors. In some instances, the construction contract requires the general contractor to purchase the builder's risk insurance and name both the owner and all the subcontractors as additional named insureds. As with other types of insurance, builder's risk insurance policies may be purchased with a deductible amount.

Builder's risk coverage may be purchased as all-risk insurance or as insurance against named perils. Coverage should be equal to the initial contract amount plus the value of any owner-furnished material and be increased by the value of any subsequent modifications in the contract amount. The policy should be written to cover property in transit and property stored at off-site locations. The policy should grant permission to occupy, allowing the building or structure to be partially occupied prior to completion, without detrimental effect to the coverage being provided. The policy should also provide coverage for the perils of earth movement (earthquake) and flood.

Most all-risk policies contain named exclusions, which often can be quite broad. Common exclusions are the cost of correcting faulty workmanship or materials, the loss due to breakdown of machinery, as well as the loss due to an error or an omission in the design of the project. Losses due to design errors or omissions are generally covered by errors and omissions insurance carried by the designer. Losses due to machinery breakdown are covered by boiler and machinery insurance.

An important aspect of builder's risk insurance is the **subrogation** clause, which grants the insurance company the right of the insured(s) to recover from the party that caused the loss. This means that the insurance company will compensate the insured(s) for the loss and then have the right to pursue action to recover its cost from the offending party. Implementation of this clause may cause major problems on a construction project if a party working on the project causes the loss. To eliminate this undesirable situation, the purchaser of the builder's risk insurance policy should identify all parties working on the project as well as the owner as additional named insureds and include a waiver of subrogation rights among all insured parties.

Bridge Insurance   **Bridge insurance** is a type of inland marine insurance that provides builder's risk type of coverage to bridge projects, which may be excluded from coverage by conventional builder's risk policies. Typical coverage includes damage to the project under construction, to any temporary structures, and to stored materials.

Boiler and Machinery Insurance   Most builder's risk and commercial general liability insurance policies exclude coverage for loss or damage due to breakdown of machinery. Consequently, **boiler and machinery insurance** may be needed to cover this uninsured risk. Such policies cover

- Financial loss due to mechanical breakdown of boilers, electrical equipment, pumps, motors, compressors, or air-conditioning equipment. Such mechanical breakdown must be accidental and cause physical damage to the covered item, necessitating its repair or replacement.
- Financial loss incurred because of mechanical failure of covered items, such as explosion of a boiler or failure of refrigeration equipment.

Property Insurance for the Construction Company's Property

Construction companies need to consider the consequences of financial loss due to loss of or damage to owned property. Depending on the type of property owned, a company may need to purchase multiple insurance policies to transfer this risk, as illustrated in Figure 3.4.

Property Insurance on Company-Owned Buildings   **Property insurance** is a first-party insurance that provides coverage for financial loss due to physical damage to a construction company's property. Such property includes offices, warehouses, fabrication facilities, and other buildings, as well as company-owned personal property, and the

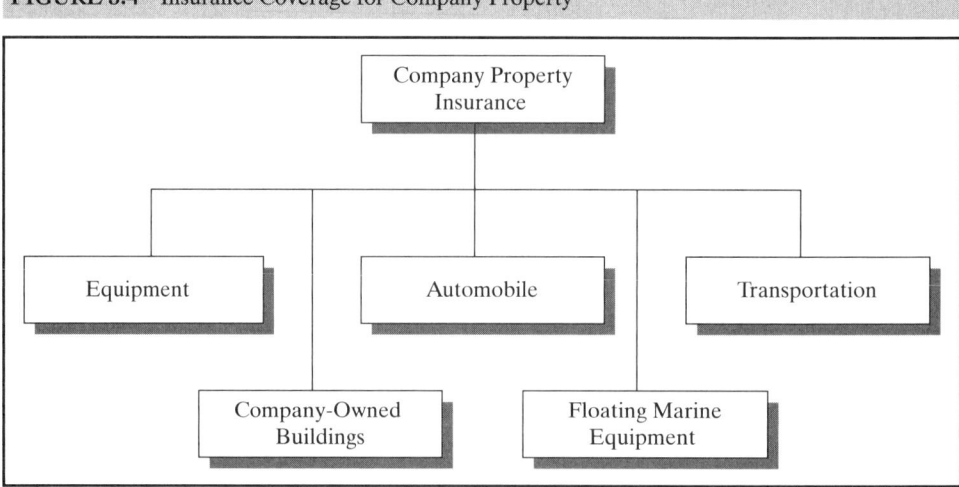

FIGURE 3.4   Insurance Coverage for Company Property

personal property of others in the company's care, custody, and control. Policies may be written to cover named perils, or may be written to cover all risks except named exclusions. A policy may provide a specific dollar value coverage for each building and an aggregate value for personal property or may be a blanket policy providing an aggregate value coverage for listed property and personal property.

Equipment Floater Insurance Equipment floater insurance is a type of inland marine coverage that protects a construction company against financial loss due to loss of or damage to

- construction equipment,
- temporary structures to support construction operations, and
- materials and supplies that will not be incorporated into a construction project.

Such policies are typically written, without reference to specific locations, to protect against financial loss due to physical loss of or external damage to equipment while on the job, in transit, or stored in the construction company's facilities. On-highway and waterborne equipment are generally excluded from coverage under these policies.

In general, equipment floater insurance policies may not cover the loss of or damage to leased or rented equipment. If such coverage is desired, it can be added for an additional fee. Policies either may only cover named perils or may provide all-risk coverage subject to listed exclusions. Such exclusions typically include damage due to equipment overload or failure to maintain and repair. This may be a major issue with owner, rented, or leased cranes. Boom overload is usually the primary risk exposure for cranes and should be included in equipment floater policies, but such coverage will increase the cost of the insurance.

Coverage may be provided on a schedule basis, with all items listed with the corresponding amount of insurance applied to each, or on a blanket basis, with an amount of insurance applied to all items together.

Transportation Floater Insurance Transportation floater insurance provides coverage for direct loss of or physical damage to property while in transit to or from a project. This type of policy covers property being transported by a public carrier, and may be purchased on a per-trip, a project, or an annual basis. This type of insurance may not be needed if equipment in transit by a common carrier is covered by an equipment floater insurance policy. Project materials in transit should be covered by a builder's risk insurance policy.

Automobile Insurance The operation of automobiles and other vehicles on highways exposes a construction company to two broad categories of risk:

- Financial loss due to the loss of or physical damage to the vehicles.
- Liability for bodily injury to others or damage to the property owned by others caused by the operation of a vehicle.

Automobile insurance is purchased by construction companies to transfer these risks to insurance companies.

The liability coverage obligates the insurer to pay all sums that the insured must legally pay as damages because of bodily injury or property damage caused by an accident, up to the limits specified in the insurance policy. The physical damage coverage obligates the insurer to cover the cost of repairing damaged vehicles, up to the limit specified in the policy and subject to any deductible. There are three forms of property damage protection:

- Damage due to collision
- Damage due to specific perils, such as hail or flying rocks
- Comprehensive coverage

Comprehensive coverage pays for damage to covered automobiles due to any cause other than collision. Automobile policies may also provide liability coverage for nonowned, rented vehicles used by employees to conduct company business. As with other insurance policies, the premiums can be reduced by accepting responsibility for a deductible amount.

Automobile insurance policies typically do not cover damage to items being transported in company vehicles. Nor do they cover liability arising from the operation of mobile equipment. This liability is covered by the commercial general liability insurance policy. Physical damage to mobile equipment is covered by equipment floater insurance.

Floating Marine Equipment Insurance   Like automobile insurance, **floating marine equipment insurance** provides protection against financial loss due to direct loss of or damage to boats, vessels, barges, and any type of floating equipment used in connection with a project. It also covers the construction company's liability for bodily injury to a third party or damage to property owned by a third party. Such floating equipment is typically excluded from coverage by both equipment floater insurance and commercial general liability insurance.

Liability Insurance

Liability insurance is needed when in the course of conducting business, a construction company happens to incur any of the following types of legal liability:

- Liability for personal injury to third parties and/or physical damage to their property.
- Liability for acts of parties for whom it is responsible, such as subcontractors.
- Liability for latent defects in completed projects.
- Liability that results from design or professional services provided.
- Liability to employees injured on the job.
- Liability for actions taken by company employees.

No one insurance policy will provide coverage for all these various liabilities, so a construction company needs to acquire a package of various liability insurance policies, as illustrated in Figure 3.5. Each type of liability insurance is described in the following sections of this chapter. Construction company leaders should consult with their insurance agents to determine the levels of coverage needed.

Commercial General Liability Insurance   Commercial general liability insurance protects the policyholder (construction company) and other named insureds against financial losses that may result from injuries or property losses sustained by third parties as a consequence of the construction firm's activities. By third parties we mean anyone except the two parties to the insurance contract, the insurer and the insured.

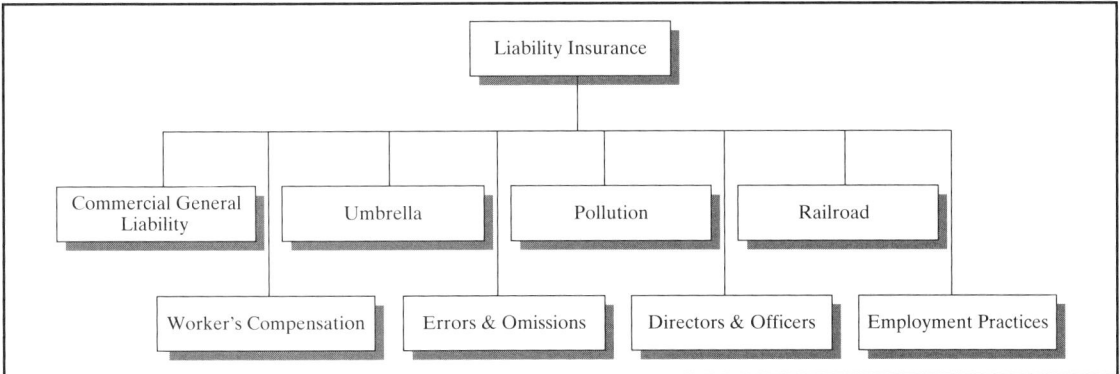

FIGURE 3.5 Liability Insurance Coverage

Covered claims include bodily injury and/or property damage caused by an accident or continuous exposure to some harmful condition. Thus water penetration into a completed building over time would be covered. Such policies do not provide any protection against financial losses incurred by the contractor as a direct consequence of its activities.

Standard policies contain three types of insuring agreements:

- Premises and operations coverage, which covers claims associated with the construction company's property or construction operations.
- Independent contractor coverage, which covers claims that result from indemnity agreements contained in most construction contracts. This contractual liability is not covered in premises and operations coverage.
- Completed operations coverage, which covers claims for damage or injury that results from the completed project years after its completion.

Premium rates are generally experience-based, which means that a construction company's claims history is considered by the insurance company when establishing a price for a desired level of coverage. A construction company's annual payroll costs are also considered by insurance companies in establishing premium costs for liability insurance.

Some of the types of claims covered under the premises and operations coverage include

- Claims from property owners near a project site for damage caused by a construction company's employees.
- Claims by third parties who are injured while visiting a construction site or the construction company's offices.
- Claims from utility companies for damages to underground pipe or cable damaged by the construction company's excavation operations.

While this coverage is broad, most commercial general liability insurance policies contain the following exclusions:

- Liability resulting from the operation of owned, rented, or leased automobiles, aircraft, and watercraft.
- Liability for damage to construction projects caused by company employees.

- Liability for injury to the construction company's employees.
- Damage caused by company employees to company-owned property.
- Liability arising from any pollution caused by company operations.
- Liability for operations occurring within 50 feet of any railroad property.
- Liability for damage to property in the care, the custody, or the control of the construction company.

Most construction contracts transfer liability between the project owner and the general contractor through indemnity agreements. Normally, the project owner requires the general contractor to hold the owner harmless. General contractors typically include similar indemnity agreements in all contracts with subcontractors. The contractual liability coverage provided by a commercial general liability insurance policy would respond to the construction company's obligation under contractual indemnity agreements. Liability that results from professional services offered by a construction company is typically excluded from coverage.

Once a construction company completes a construction project and turns it over to the owner, the construction company can be held liable for damage or injury resulting from latent defects for years in the future. While the project owner accepts responsibility for any defects that are readily discoverable by visual inspection upon acceptance of the completed project, the construction contractor may be held liable for any hidden unsafe conditions or latent defects.

Commercial liability insurance policies are written in one of two forms. They may be either occurrence-based policies or claims-made policies. An **occurrence-based policy** provides coverage regardless of when the claim is made, provided the insured occurrence took place during the policy period. For example, if an automobile was damaged because a crane dropped a piece of steel in 2005, and the claim was not submitted until 2006, the contractor's insurance company would still cover the claim as long as a valid insurance policy was in effect in 2005, whether or not the policy was in effect in 2006. A **claims-based policy** covers only claims made during the policy period, irrespective of when the insured occurrence took place. From the perspective of a construction company, an occurrence-based policy is preferred and is the most commonly used form.

Another issue to consider is whether or not the cost of defense is included in the policy coverage limits. If the cost of defense is excluded or is in addition to the policy limits, the cost the insurance company incurs in defending a claim does not reduce the value of the coverage.

Umbrella Excess Liability Insurance An **umbrella excess liability insurance** policy is designed to provide a construction company with high limits of liability coverage when the coverage provided by other liability insurance policies has been exhausted. This means that the umbrella policy provides excess liability in addition to the limits provided by other liability insurance policies. An umbrella policy is intended to provide additional liability insurance coverage in the event of a catastrophic loss, and the umbrella underwriter will want to ensure that there is valid and adequate liability insurance coverage provided by the policies underneath the umbrella policy. This is illustrated in Figure 3.6.

Unlike other forms of commercial insurance, there is no standard umbrella insurance form. Each insurance company drafts its own form. Coverage is sold on either an occurrence-based or claims-made basis. Construction company leaders in consultation

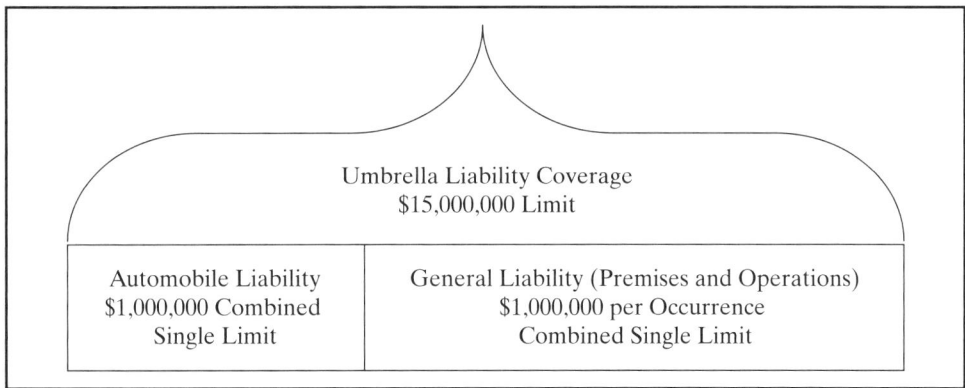

FIGURE 3.6 Umbrella Excess Liability Insurance Coverage

with their insurance agents should carefully review the policy language to ensure that they understand which liabilities are covered and which are excluded.

Railroad Liability Insurance There are exclusions in most commercial general liability policies for bodily injury and/or property damage that result from construction work performed within 50 feet of railroads or on a railroad easement. Railroads typically require construction firms to hold the railroads harmless for injury or damage that occurs as a result of the construction company's work. Since coverage is not provided by other liability insurance policies, the construction firm should purchase **railroad liability insurance** to cover bodily injury or property damage claims that may result from its work near railroads.

Errors and Omissions Insurance **Errors and omissions insurance** is a professional liability insurance that covers the professional negligence of design professionals. Such policies are needed, because most commercial general liability and umbrella excess liability insurance policies have professional liability exclusions. Some project owners are requiring professional construction managers to carry similar insurance. A construction firm that wishes to compete for design-build projects may be required to carry errors and omissions insurance in order to qualify for the project. Most policies cover both legal defense costs as well as any legal judgments against the insured, including court costs, up to the limits described in the policy.

The policies pay on behalf of the insured all sums that the insured is legally obligated to pay as damages by reason of any negligent act, error, or omission. This type of insurance protects the insured from claims alleging that the insured has been negligent in the performance of professional obligations. Such policies also cover liability for failure to act as a competent professional. Most policies exclude claims arising out of the failure to complete drawings or specifications in a timely manner.

Most errors and omissions insurance policies are written on a claims-made basis, which means that claims must be made during the period covered by the policy, irrespective of when the alleged negligent act occurred. Professional liability insurance policies typically include a deductible amount that is applied to each claim. When professional liability insurance is obtained, it is recommended that coverage include liability for any professional work performed by subcontractors on behalf of the construction company. This is especially important if any subcontractors perform design services on behalf of the general contractor.

Pollution Liability Insurance **Pollution liability insurance** provides coverage for legal liability relating to pollution conditions. Construction companies need to purchase this coverage, because pollution liability is typically excluded from coverage by commercial general liability insurance policies. This coverage is usually divided into two basic parts:

- Coverage for pollution claims made against the insured by third parties, whether on, under, or beyond the boundaries of the insured's property. Coverage includes clean-up, bodily injury, and property damage.
- Coverage for clean-up of pollution conditions discovered on the insured's property.

Pollution liability insurance policies are typically written on a claims-made basis, meaning that in order for the claim to be covered, it must be discovered and reported to the insurance company during the life of the policy. Punitive damages, civil fines, or penalties are typically excluded from coverage by pollution liability insurance policies.

Employment Practices Liability Insurance **Employment practices liability insurance** provides protection for a construction company against claims made by employees, former employees, or potential employees. These policies cover claims relating to discrimination, wrongful termination of employment, sexual harassment, and other employment-related allegations. Most employment practices liability insurance policies include a duty to defend as well as to claim settlement. The cost of coverage depends on the number of employees the company has and the company's history regarding past litigation over employment practices. To reduce premium cost, many construction companies purchase this type of insurance with a deductible. The best strategy for avoiding these types of claims is to have a formal and effective written manual of employment practices.

Directors and Officers Liability Insurance **Directors and officers liability insurance** provides company directors and officers protection from personal liability and financial loss arising out of acts committed or allegedly committed in their capacity as company directors or officers. The policy provides indemnification to the directors and officers for any legal liability incurred as a result of any action, suit, or proceeding related to their job performance. Most directors and officers liability insurance policies are written on a claims-made basis.

Worker's Compensation Insurance **Worker's compensation insurance** is a no-fault insurance that is mandated by state law, in that the employer cannot deny a claim by an insured employee on the basis that the employee was negligent in causing the injury. Likewise, the employee cannot sue the employer on the basis that the employer's negligence contributed to the work-related injury or disease.

The dollar value of the construction firm's liability is usually established by statute. The benefits provided are generally medical benefits, loss of earnings, and retraining if the employee is unable to perform the duties of the current position. In some states, the insurance is offered by monopolistic state agencies, while in others it is available from insurance carriers. Construction firms that meet state requirements may choose to be self-insuring. The essence of a worker's compensation insurance contract is that the insurer agrees, for a price (premium), to assume the liability imposed on the insured construction company by the worker's compensation statutes of the state named in the policy.

Under the worker's compensation statutes of the various states, the employer's liability is limited to a specific benefit level when an employee is injured or killed on the job. The benefits are automatic and cannot be appealed by the employer. Because worker's compensation insurance is a type of no-fault insurance, a construction company cannot be sued by its employees or their heirs for additional compensation as a result of injury or death.

Construction projects involving operations on or over navigable waterways come under the jurisdiction of two federal statutes. These statutes are the U.S. Longshoremen's and Harbor Worker's Act and the Jones Act. Benefits under these laws may vary from those provided by state insurance plans.

Worker's compensation premiums are based on the construction company's payroll and differ for each employee work classification based on the degree of injury risk being assumed by a specific craft. For example, the premium rate for a steel worker is typically higher than that for a plumber. The actual rates are determined annually by a state government agency based on the past losses attributed to each work classification.

Worker's compensation premium rates have two components. The first is the base rate, which is applied to each $100 of direct employee compensation. Thus the base premium for each work classification per pay period is

$$\text{Base Premium} = \frac{(\text{Direct Wages})(\text{Base Rate})}{\$100}$$

The second component is a premium modifier, known as the **experience modification ratio**, EMR, which is based on the company's claim history in the oldest three of the past four years. It is determined as follows:

$$\text{EMR} = \frac{\text{Aggregate Claims over 3-year Period}}{\text{Expected Claims over 3-year Period}}$$

Companies with good safety records typically have EMRs below 1.0, while the rates for companies with many claims are generally above 1.0. The actual premium rates paid by a company are

$$\text{Modified Premiums} = (\text{Base Premiums})(\text{EMR})$$

For example, let's assume that the base rate for an employee is $3.50 and that the company EMR is 0.8. If the employee earns $1,200 per pay period (week), the weekly worker's compensation premium rate would be

$$\frac{(\$1,200)(\$3.50)(0.8)}{\$100} \text{ or } \$33.60$$

Thus, a company's labor cost on a project is directly influenced by its safety record, as defined by its EMR.

Worker's compensation insurance may also be sold on a retrospective basis. The premium is calculated and paid based on the base premium rates. The insurance company or state agency periodically evaluates the construction company's claims under the policy. If the claims costs exceed the premiums paid, the construction company is required to pay the additional premium. If the claims costs are lower than the premium, the insurer refunds the difference to the construction company. Retrospective rating plans are often referred to as cost-plus insurance.

Business Insurance

Key-Person Insurance Key-person insurance is life insurance obtained by the construction company on company principals. These policies provide protection to the company against financial losses that may result from the untimely death of a principal. They simply are large life insurance policies that are owned by the company. The benefits from such policies are generally used to finance buy/sell agreements that the deceased made with the company as a part of its succession plan. Without such an infusion of capital, the company may be unable to purchase the financial interests of the deceased without resorting to the sale of company assets.

Subcontractor Insurance

By the terms of its subcontracts, general contractors typically require each subcontractor to provide and maintain certain levels of insurance coverage. Usually, the coverage limits required of subcontractors are the same as those required of the general contractor by the project owner. Many general contractors also require subcontractors to list the general contractors as additional insured on the subcontractors' liability insurance policies. This is an important part of a general contractor's overall risk management strategy.

Insurance Certificate

Under the terms and conditions of most construction contracts, the contractor is required to submit a certificate of insurance, similar to the one illustrated in Figure 3.7. These certificates are issued by insurance companies to certify that the named insureds have in force the coverage listed. As a part of its risk management strategy, general contractors should require similar certificates from all subcontractors before allowing them to initiate work on a construction project.

BONDING

Surety Relationship

A surety is a party that assumes liability for the performance of another party. A **surety bond** is a three-party contract, as illustrated in Figure 3.8. The construction company, known as the **principal**, purchases a bond from a surety, guaranteeing to the project owner, the **obligee**, that the construction firm will perform all its obligations as specified in the construction contract. Should the construction company default on its contractual obligations, the surety agrees to step in and perform the contractual obligations or to compensate the obligee for any additional costs up to the **penal sum** of the bond. The construction company pays the surety a fee or premium in exchange for the bond guaranteeing performance to the obligee. Most surety companies in the United States are divisions of major insurance companies.

The surety, in effect, is assuming the obligation of the principal (contractor) in complying with the terms of the construction contract. In many respects, the surety

FIGURE 3.7 Insurance Certificate

CERTIFICATE OF INSURANCE		
Producer (Insurance Broker) *Cascade Insurance Agency*		THIS CERTIFICATE IS ISSUED AS A MATTER OF INFORMATION ONLY AND CONFERS NO RIGHTS UPON THE CERTIFICATE HOLDER. THIS CERTIFICATE DOES NOT AMED, EXTEND OR ALTER THE COVERAGE AFFORDED BY THE POLICIES BELOW.

	COMPANIES AFFORDING COVERAGE	
Insured *Western States Construction Company*	INSURANCE COMPANY	BEST RATING
	Company Letter A — *National Insurance*	*AAA*
	Company Letter B — *National Insurance*	*AAA*
	Company Letter C — *National Insurance*	*AAA*
	Company Letter D	

COVERAGES

THIS IS TO CERTIFY THAT THE INSURANCE POLICIES LISTED BELOW HAVE BEEN ISSUED TO THE INSURED NAMED ABOVE FOR THE POLICY PERIOD INDICATED. NOTWITHSTANDING ANY REQUIREMENT, TERM, OR CONDITION OF ANY CONTRACT OR OTHER DOCUMENT WITH RESPECT TO WHICH THIS CERTIFICATE MAY BE ISSUED OR MAY PERTAIN, THE INSURANCE AFFORDED BY THE POLICIES DESCRIBED BELOW IS SUBJECT TO ALL THE TERMS, EXCLUSIONS, AND CONDITIONS OF SUCH POLICIES. LIMITS MAY HAVE BEEN REDUCED BY PAID CLAIMS.

Co. Ltr.	Type of Insurance	Policy Number	Policy Effective Date (M/D/Y)	Policy Expiration Date (M/D/Y)	Limits	

A	**General Liability**
	X Commercial General Liability
	–
	–
		27ABC649	*10/1/05*	*9/30/06*	General Aggregate
					Products/Comp
					  Ops Agg
					Pers. & Adv. Injury
					Each Occurrence
					Fire Damage
					  (any one fire)
					Med. Expenses
					  (any one person)
						$2,000,000
						
						$2,000,000
						$2,000,000
						$2,000,000
						
						$1,000,000
						
						$100,000

B	**Automobile Liability**
	X Any Automobile
	–
		27ABC659	*10/1/05*	*9/30/06*	Combined Single
					  Limit
					Bodily Injury
					  (per person)
					Bodily Injury
					  (per accident)
					Property Damage
						
						$1,000,000
						
						$500,000
						
						$1,000,000
						$500,000

(continued)

FIGURE 3.7 (continued)

C	**Excess Liability** X Umbrella Form –	27ABC789	10/1/05	9/30/06	Each Occurrence Aggregate	$10,000,000 $20,000,000
D	**Worker's Compensation**				Each Accident Disease Policy Limit Disease (each employee)	$ $ $

PROJECT NAME FOR ALL OPERATIONS:

Northwest Development Company is an additional insured per attached endorsement. The liability insurance referred to in this certificate is primary and non-contributory to any insurance carried by *Western States Construction Company* per attached endorsement.

CERTIFICATE HOLDER	**CANCELLATION**
Northwest Development Company	SHOULD ANY OF THE ABOVE DESCRIBED POLICIES BE CANCELED BEFORE THE EXPIRATION DATE LISTED, THE ISSUING COMPANY WILL PROVIDE 7 DAYS' WRITTEN NOTICE TO CERTIFICATE HOLDER NAMED TO THE LEFT. **AUTHORIZED REPRESENTATIVE** *Henry Jones*

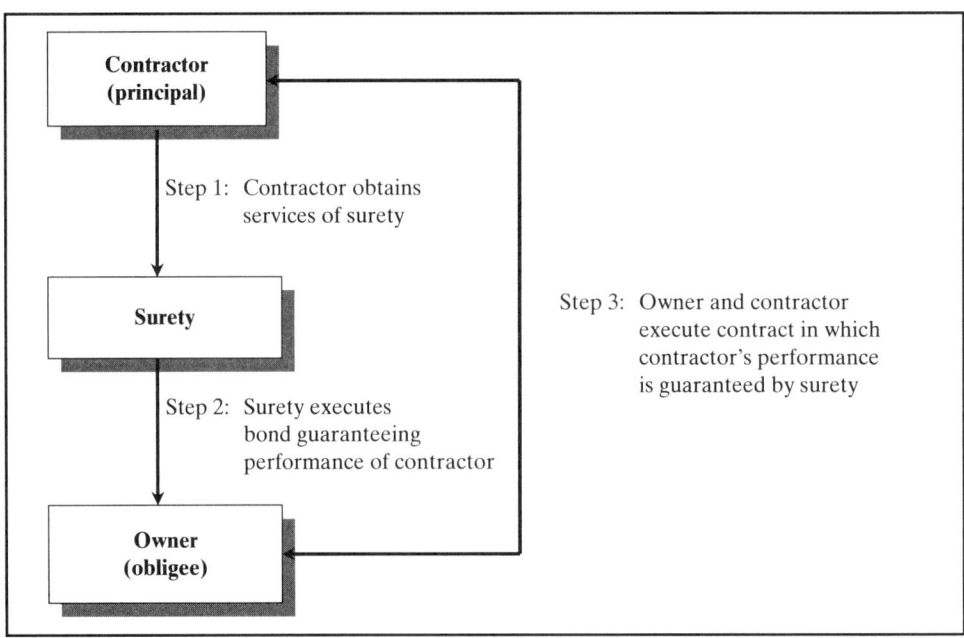

FIGURE 3.8 Surety Relationship

is granting credit to the principal by allowing the surety's financial resources to guarantee the principal's performance of its contract requirements. Unlike bank credit, the surety does not actually loan money to the contractor. In exchange for the credit provided by the bond, the construction company pays a fee, or premium, which generally averages between 1 and 3 percent of the bond value. The amount of the premium charged depends on such factors as

- the total project cost,
- the project duration,
- the type of project, and
- the construction firm's financial condition.

A surety bond is not an insurance policy, as illustrated in Figure 3.9. The surety is a guarantor of the performance of the principal, not an insurer of the obligee. The surety company guarantees the obligee that the principal will perform its required obligations and that if not performed, the surety company will do so. A surety agreement is a three-party agreement, while an insurance contract is a two-party agreement in which the insurance company agrees to assume those risks of the insured's that are specified in the insurance policy. Insurance protects a party from the risk of loss, while a surety bond guarantees the performance of a contractual responsibility.

In addition to the premium, the construction company is generally required to sign a separate contract, known as an indemnity agreement, with the surety company. This **indemnity agreement** provides that the construction company will pay the surety back for any losses that the surety incurs in making good on the guarantee to the obligee. Most sureties require indemnification agreements from the construction company, which incorporate any and all bonds issued on behalf of the company. If the principal is a corporation, the corporate officers may be required to sign as personal indemnitors on behalf of the corporation. The essence of their commitment is to cover any losses

FIGURE 3.9 Comparison of Surety Bond with Insurance

Contract Surety	**Insurance**
Type Agreement	
3-party agreement	2-party agreement
Term of Obligation	
indefinite—until contractual obligation is performed	stated policy period
Coverage	
performance	named perils
Indemnity Available to Surety or Insurance Company	
indemnity agreement signed by company officers on behalf of principal	none

the surety may incur by providing bonds to the corporation. A sample indemnity agreement is shown in Figure 3.10. While insurance premium rates are established based on anticipated losses, sureties expect no losses because they are selecting to bond construction firms based on the firm's credit strength and construction expertise. Construction company executives retain the economic risk of contract default by signing indemnity agreements that indemnify the surety and hold it harmless for all expenses that the surety incurs as a result of invocation of the bond.

The demand for contract bonds usually comes from one of the following sources:

- Statutory requirement for public (government) projects.
- Project owner requires as part of its risk management strategy.
- General contractor requires from subcontractors either as a part of its risk management strategy or at the request of the contractor's surety.

Thus the ability to obtain construction bonding may be critical to a construction company's ability to pursue certain types of projects.

Bonding Agents

As with insurance, construction companies purchase bonds through bonding brokers or agents, who also are known as *surety bond producers*. The **bonding agent** acts as a representative of the construction company in locating the most advantageous bonding arrangement with an underwriting company (surety). The determination of the amount of bond coverage to provide for a construction company is made by the underwriter, who is an employee of the surety company. As with insurance agents, a construction company typically establishes an enduring relationship with a single bonding agent.

When selecting a bonding agent, a construction company should seek a firm that

- Has experience with construction and has current clients performing similar types of construction work.
- Has a reputation for honesty and integrity.
- Has access to multiple surety markets.
- Has an awareness of local, regional, and national construction markets.
- Is actively involved in construction and surety industry associations.
- Is a good personality match with the business culture of the construction firm.

The construction company leaders must feel comfortable with the bonding agent and confident in his or her ability to represent the company's interests to the surety.

If necessary, a construction company should not be reluctant to change agents, but generally it is best practice to use an agent who is very familiar with the construction firm. The bonding agent helps the construction company establish a bond program, which revolves around two figures, (1) the largest size project that the company should undertake and (2) the total amount of bonded work outstanding at any one time. The bonding agent should periodically visit the construction company's projects and should arrange periodic meetings between the construction firm and the surety, preferably at least once per year.

Types of Bonds

A surety bond is a contract that describes the conditions and obligations of the surety. Standard forms for some types of construction bonds are published by the General

General Agreement of Indemnity for Contractors

This AGREEMENT is made by the Undersigned in favor of the American Surety Company for the purpose of indemnifying them from all loss and expense in connection with any Bonds for which the American Surety Company becomes Surety for the Western States Construction Company. In consideration of the execution of any such Bonds issued by the American Surety Company, the Undersigned, jointly and severally, agree as follows:

INDEMNITY TO SURETY: Undersigned agree to pay Surety upon demand:

1. All loss, costs, and expenses of whatsoever kind and nature, including court costs, reasonable attorney fees, consultant fees, investigative costs, and any other losses, costs, or expenses incurred by Surety by reason of having executed any Bond, or incurred by it on account of any default under this Agreement by any of the Undersigned.
2. An amount sufficient to discharge any claim made against Surety on any Bond.
3. Any original, additional, or renewal premium due for any Bond.

SURETY'S REMEDIES IN EVENT OF DEFAULT: In event of default by Western States Construction Company, Surety shall have the right, at its sole discretion, to:

1. Take possession of the work under any and all contract and to arrange for its completion by others or by the Obligee of any Bond.
2. Take possession of Contractor's or any of the Undersigned's equipment, materials, and supplies at the site of work, or elsewhere, if needed for the prosecution of the work, without payment of rental for such use.
3. Guarantee a loan to Contractor of such money as Surety sees fit for the purpose of completing any contract.
4. Take such other action as Surety shall deem necessary to fulfill its obligations under any Bond.

SECURITY TO SURETY: As collateral security to Surety for the agreement of the Undersigned to repay all losses and expenses to Surety, the Undersigned assigns to Surety, as of the date of execution of any Bond, all rights of the Contractor in:

1. Any contract or modification thereof.
2. Any subcontract or purchase order.
3. Monies due or to become due to Contractor on any contract.
4. All monies, bank accounts, and deposits.
5. All equipment, machinery, and tools.
6. All securities and investments.

TERMINATION: This agreement is a continuing obligation of the Undersigned as long as they have Bonds issued by Surety in effect. Once all bonds lapse, the agreement may be terminated by written notification.
    This agreement is effective this fifteenth day of December 2005.

WITNESS: INDEMNITOR:
Kevin T. Brown   *James B. Smith*
Kevin T. Brown   James B. Smith, President
                                  Western States Construction Company

FIGURE 3.10 Indemnity Agreement

Services Administration, the American Institute of Architects, and the Associated General Contractors of America, while some surety companies use their own forms. Now let's look at the types of bonds that are most common in the construction industry.

Bid Bond

A **bid bond** is a guarantee that the successful bidder will furnish a performance bond and any other contractual requirements and enter into a construction contract for the price contained in its bid. If the contractor fails to provide the required bonds and/or sign the contract, the bond stipulates that the surety will pay for the additional cost incurred by awarding the contract to the next low bidder. The bid bond of the low bidder becomes null and void once the performance bond is provided and the contract is signed. A bid bond generally states a specified penal sum, which is usually 5 to 20 percent of the value of the total bid. The fee for issuing a bid bond is usually nominal, or may be free. An example of a bid bond is shown in Figure 3.11.

FIGURE 3.11 Bid Bond

American Surety Company
126 William Street
New York, New York 10038

Bond No. BQ-157

KNOW ALL BY THESE PRESENTS, That we, Western States Construction Company of Seattle, Washington (herein called the Principal) as Principal and American Surety Company, a New York Corporation of New York, New York (herein called the Surety) as Surety are held and firmly bound unto the City of Seattle (herein called the Obligee) in the penal sum of nine hundred eighty-five thousand dollars ($985,000) for the payment of which the Principal and the Surety bind themselves, their heirs, executors, administrators, successors and assigns, jointly and severally, firmly by these presents.

THE CONDITION OF THIS OBLIGATION IS SUCH, That WHEREAS the Principal has submitted a bid to the Obligee on a contract for construction of the Sand Point Library.

NOW, THEREFORE, If the said contract be timely awarded to the Principal and if the Principal shall, within such time as may be specified, enter into the contract and give Bond with surety acceptable to the Obligee for performance of said contract, then this obligation shall be void; otherwise to remain in full force and effect.

Signed and sealed this fifth day of January, 2006.

Harry P. Jones
(Witness)

James B. Smith
James B. Smith, President
Western States Construction Company
(Principal)

Samuel T. White
(Witness)

Terry L. Johnson
Terry L. Johnson, Attorney-in-Fact
American Surety Company
(Surety)

Performance Bond

A **performance bond** is a guarantee that the construction contractor will perform according to the terms of the construction contract. A performance bond usually states a specified penal sum to limit the liability of the surety, which is often 100 percent of the contract value. The standard form used by the General Services Administration and other federal agencies provides that the bond will apply to all contract modifications and the life of the bond includes all extensions of contract time. Performance bonds typically contain a provision that in the event of a default by the construction contractor, the surety has the choice of either completing the construction contract or paying the project owner the additional costs incurred in completing the contract, up to the value of the bond. Bond coverage is for the life of the contract which includes the one-year warranty period. Bond premiums are based on the value of the coverage. The rate per $1,000 coverage decreases as the size of the bond increases. An example of a performance bond is shown in Figure 3.12.

FIGURE 3.12 Performance Bond

American Surety Company
126 William Street
New York, New York 10038

Bond No. BP-157

KNOW ALL BY THESE PRESENTS, That we, Western States Construction Company of Seattle, Washington (herein called the Principal) as Principal and American Surety Company, a New York Corporation of New York, New York (herein called the Surety) as Surety are held and firmly bound unto the City of Seattle (herein called the Obligee) in the penal sum of nine million, eight hundred fifty thousand dollars ($9,850,000) for the payment of which the Principal and the Surety bind themselves, their heirs, executors, administrators, successors and assigns, jointly and severally, firmly by these presents.

THE CONDITION OF THIS OBLIGATION IS SUCH, That WHEREAS the Principal has entered into a written contract with said Obligee, dated January 16, 2006, for the Sand Point Library in accordance with the terms and conditions of said contract, which is hereby referred to and made a part hereof as if fully set forth herein.

NOW, THEREFORE, If the Principal shall promptly and faithfully perform said contract and reimburse the Obligee for all loss and damage sustained by reason of failure or default on the part of the Principal, then this obligation shall be void; otherwise to remain in full force and effect.

Signed and sealed this sixteenth day of January, 2006.

Harry P. Jones
(Witness)

James B. Smith
James B. Smith, President
Western States Construction Company
(Principal)

Samuel T. White
(Witness)

Terry L. Johnson
Terry L. Johnson, Attorney-in-Fact
American Surety Company
(Surety)

Labor and Material Bond

A **labor and material bond** (sometime called a payment bond) is a guarantee that the construction contractor will compensate any entity that provided labor or materials for the construction project. Construction projects are subject to **mechanic's lien** laws, which enable subcontractors and suppliers to file legal claims, called liens, against the project property if they are not compensated for their labor or materials. Labor and material bonds assure that the surety will pay these claims, in the event that the construction contractor refuses to pay. Coverage is usually the same as that required for payment bonds. Like performance bonds, premiums are based on the value of the coverage. An example of a labor and material bond is shown in Figure 3.13.

Maintenance Bond

A **maintenance bond** (sometimes called a warranty bond) is a guarantee that the construction contractor will return to the completed project and repair or replace any

FIGURE 3.13 Labor and Material Bond

American Surety Company
126 William Street
New York, New York 10038

Bond No. LM-157

KNOW ALL BY THESE PRESENTS, That we, Western States Construction Company of Seattle, Washington (herein called the Principal) as Principal and American Surety Company, a New York Corporation of New York, New York (herein called the Surety) as Surety are held and firmly bound unto the City of Seattle (herein called the Obligee) in the penal sum of nine million, eight hundred fifty thousand dollars ($9,850,000) for the payment of which the Principal and the Surety bind themselves, their heirs, executors, administrators, successors and assigns, jointly and severally, firmly by these presents.

THE CONDITION OF THIS OBLIGATION IS SUCH, That WHEREAS the Principal has entered into a written contract with said Obligee, dated January 16, 2006, for the Sand Point Library in accordance with the terms and conditions of said contract, which is hereby referred to and made a part hereof as if fully set forth herein.

NOW, THEREFORE, If the Principal shall promptly and faithfully make payment to all Claimants for all labor and material used or reasonably required for use in the performance of the contract, then this obligation shall be void; otherwise to remain in full force and effect.

Signed and sealed this sixteenth day of January, 2006.

Harry P. Jones
(Witness)

Samuel T. White
(Witness)

James B. Smith
James B. Smith, President
Western States Construction Company
(Principal)

Terry L. Johnson
Terry L. Johnson, Attorney-in-Fact
American Surety Company
(Surety)

defective or inferior materials or workmanship during the specified warranty period. Maintenance bonds may be required for components with warranties that exceed the normal one-year warranty. An example might be a ten-year warranty on a roof. The bond value may be based on a percentage of the contract value, such as 20 percent, or based on the value of a specific component, such as a roof. An example of a maintenance bond is shown in Figure 3.14.

Supply Bond

A **supply bond** is a guarantee to the construction firm that a material supplier will comply with the terms of a purchase order or supply contract. This includes

FIGURE 3.14 Maintenance Bond

**American Surety Company
126 William Street
New York, New York 10038**

Bond No. MT-145

KNOW ALL BY THESE PRESENTS, That we, Western States Construction Company of Seattle, Washington (herein called the Principal) as Principal and American Surety Company, a New York Corporation of New York, New York (herein called the Surety) as Surety are held and firmly bound unto the Emerald City Development Corporation (herein called the Obligee) in the penal sum of eight hundred fifty thousand dollars ($850,000) for the payment of which the Principal and the Surety bind themselves, their heirs, executors, administrators, successors and assigns, jointly and severally, firmly by these presents.

THE CONDITION OF THIS OBLIGATION IS SUCH, That WHEREAS the Principal has entered into a written contract with said Obligee, dated March 18, 2005, for the Cascade Medical Office Building that provides that the Principal will furnish a bond conditioned to guarantee the roof of said facility for a period of ten years against all defects in workmanship and materials which may become apparent during said period, said contract, which is hereby referred to and made a part hereof as if fully set forth herein, and WHEREAS said contract was completed on January 4, 2006.

NOW, THEREFORE, If the Principal shall promptly and faithfully make good, repair, and replace at its own expense any defects discovered prior to January 4, 2016 in work done or material furnished under said contract relating to the roof, then this obligation shall be void; otherwise to remain in full force and effect.

Signed and sealed this fourth day of January, 2006.

Harry P. Jones	*James B. Smith*
(Witness)	James B. Smith, President
	Western States Construction Company
	(Principal)
Samuel T. White	*Terry L. Johnson*
(Witness)	Terry L. Johnson, Attorney-in-Fact
	American Surety Company
	(Surety)

both the correct quantity and quality of ordered materials and the delivery by a specified date. The liability of the surety is for any cost incurred by the construction company due to late delivery of the ordered materials. An example of a supply bond is shown in Figure 3.15.

Underwriting Considerations

The primary function of the surety is to prequalify and issue bonds on behalf of construction companies deemed capable of performing the construction contract. Because of this, sureties prequalify a construction company through an in-depth examination of its business operation. This process involves determining the liabilities

FIGURE 3.15 Supply Bond

American Surety Company
126 William Street
New York, New York 10038

Bond No. SP-126

KNOW ALL BY THESE PRESENTS, That we, Northwest Electrical Supply Company of Seattle, Washington (herein called the Principal) as Principal and American Surety Company, a New York Corporation of New York, New York (herein called the Surety) as Surety are held and firmly bound unto the Western States Construction Company (herein called the Obligee) in the penal sum of ten million, five hundred thousand dollars ($10,500,000) for the payment of which the Principal and the Surety bind themselves, their heirs, executors, administrators, successors and assigns, jointly and severally, firmly by these presents.

THE CONDITION OF THIS OBLIGATION IS SUCH, That WHEREAS the Principal has entered into a written contract with said Obligee, dated November 15, 2005, for purchase and delivery of two (2) generator units for the North Shore Power Plant in accordance with the terms and conditions of said contract, which is hereby referred to and made a part hereof as if fully set forth herein.

NOW, THEREFORE, If the Principal shall promptly and faithfully perform said contract and reimburse the Obligee for all loss and damage sustained by reason of failure or default on the part of the Principal, then this obligation shall be void; otherwise to remain in full force and effect.

Signed and sealed this fifteenth day of November, 2005.

Karen B. Taylor
(Witness)

Henry L. Jamison
Henry L. Jamison, President
Northwest Electrical Supply Company
(Principal)

Samuel T. White
(Witness)

Terry L. Johnson
Terry L. Johnson, Attorney-in-Fact
American Surety Company
(Surety)

of the company and investigating the background, the capabilities, and the financial condition of the firm. This will include

- resumes of key personnel,
- company organization chart,
- company business plan,
- company continuity of operation plan,
- company tax returns,
- bank statements,
- audited financial statements,
- personal financial statements of company owners/officers,
- list of owned equipment,
- list of major subcontractors,
- list of major suppliers,
- work in progress, and
- list of current and past customers.

The purpose of this research is to allow the surety company to determine how much risk it is willing to assume relative to guaranteeing the construction firm's contractual performance. The surety underwriter typically works with the construction firm's bonding agent to gather and evaluate the information that will be used in making the underwriting decision.

The bonding agent evaluates the construction company's ability to perform bonded work, recommends a surety company, and facilitates communications between the construction company and the surety underwriter. The agent compiles information regarding the construction company's financial condition and work experience. Most bonding agents require construction companies to complete a questionnaire similar to the one illustrated in Figure 3.16 as a part of this process. If, in the agent's opinion, the construction firm is capable of performing such work satisfactorily, the agent submits the collected information to a surety underwriter.

The underwriter then analyzes the information and establishes the maximum work program to be bonded. To do this, the underwriter wants to know everything possible about the construction company. This includes the company's background, the current situation, and its future plans. The underwriter's goal is to understand the strengths and weaknesses of the construction firm. This understanding is needed to prequalify the company, evaluating not only what the company can do if everything goes well, but also what resources the company has in order to absorb losses if something goes wrong.

The primary factors considered by surety underwriters in determining the extent of surety coverage (maximum aggregate value of issued bonds) that can be approved for a construction firm are character, capacity, and capital. The objective of the surety company is to extend bond coverage selectively to construction firms that present little risk of financial loss to the surety, even though covered firms will execute indemnity agreements.

Character is the reputation of the company and its leadership. This is extremely important, because the surety company is being asked to guarantee the performance of these individuals. It involves considering a construction firm's past performance and how well the company fulfills its obligations, including treatment of subcontractors

FIGURE 3.16 Contractor Questionnaire

Thomas White Agency
1700 Pike Street, Suite 500
Seattle, Washington 98101
CONTRACTOR QUESTIONNAIRE

1. Name of Firm: _____
2. Address: _____ 3. Fiscal Year End _____
   _____   _____   _____
   (city)   (state)   (zip)
4. Phone: _____ 5. Contracting Specialty: _____
6. Contract Person: _____ 7. Title: _____
8. Year Business Started: _____ 9. Type of Business: [ ] Corp. [ ] Part. [ ] Prop. [ ] Other
10. State of Incorporation: _____ 11. Area of Operation: _____
12. List corporate officers, partners, or proprietors of firm:

Name	Year of Birth	Position	Percent Owned	Name of Spouse
a.				
b.				
c.				
d.				

13. Will the above-named individuals and spouses personally indemnify Surety? [ ] Yes [ ] No   If no, explain: _____
14. Is there a buy/sell agreement among the owners of the business? [ ] Yes [ ] No
15. Is this agreement funded by life insurance? [ ] Yes [ ] No
16. How many people does your firm employ? _____ 17. How many work crews? _____
18. Has your firm or any of its principals ever petitioned for bankruptcy, failed in business, or defaulted so as to cause a loss to a Surety? [ ] Yes [ ] No   If yes, explain: _____
19. Is your firm or any of its owners currently involved in any litigation? [ ] Yes [ ] No   If yes, explain: _____
20. Typical percentage of the firm's work: Government Agencies _____ Private Owners _____
21. What percentage of the firm's work is normally subcontracted: _____
22. Are bonds required of subcontractors? [ ] Yes [ ] No
23. What trades do you normally subcontract? _____
24. What is the largest amount of uncompleted work on hand at one time in the past?
    Amount: $_____ Year: _____
25. What is the largest job you expect during the next year? $_____
26. What is the largest uncompleted work program expected during the next year? $_____

(continued)

27. What is your expected annual volume next year? $_____
28. What trades do you normally undertake with your own forces? _____
29. Do you lease equipment? [ ] Yes [ ] No   30. Type of lease: _____
31. What are the terms of the lease? _____
32. Name of your Certified Public Accountant: _____
    Address: _____
    Telephone: _____ Contact Person: _____
33. On what basis are taxes paid? [ ] Cash [ ] Completed job [ ] Accrual [ ] % of completion
34. On what basis are financial statements prepared? [ ] Cash [ ] Completed job [ ] Accrual [ ] % of completion
35. On what level of assurance are financial statements prepared? [ ] CPA audit [ ] Review [ ] Compilation
36. How often are financial statements prepared? [ ] Annually [ ] Semi-annually [ ] Quarterly [ ] Monthly
37. Do you have a full-time accountant on staff? [ ] Yes [ ] No   38. Yrs. Experience: _____
39. Are job cost records kept? [ ] Yes [ ] No
40. How often reviewed? _____   41. How often updated? _____
42. Name of your bank: _____ Contact Person: _____
    Address: _____ Telephone: _____
43. Amount of line of credit: $_____ Expiration date: _____
    Interest rate: _____
44. Is your firm union? [ ] Yes [ ] No
45. What is firm's Dun & Bradstreet Number? _____   46. D & B Rating: _____
46. Previous bonding companies:

Name	Reason for Leaving
a. _____	_____
b. _____	_____
c. _____	_____

47. List three of your largest contracts:

Job Name	Contract Price	Gross Profit	Completion Date	Bonded?
a. _____	$_____	$_____	_____	[ ] Yes [ ] No
Owner: _____		Designer: _____		
b. _____	$_____	$_____	_____	[ ] Yes [ ] No
Owner: _____		Designer: _____		
c. _____	$_____	$_____	_____	[ ] Yes [ ] No
Owner: _____		Designer: _____		

(*continued*)

FIGURE 3.16 (*continued*)

48. List three of your major suppliers:

	Name	Address	Telephone	Contact
a.	_____	_____	_____	_____
b.	_____	_____	_____	_____
c.	_____	_____	_____	_____

49. List three subcontractors (or contractors if you are a subcontractor) with whom you do business:

	Name	Address	Telephone	Contact
a.	_____	_____	_____	_____
b.	_____	_____	_____	_____
c.	_____	_____	_____	_____

50. List three architects with whom you have done business:

	Name	Address	Telephone	Contact
a.	_____	_____	_____	_____
b.	_____	_____	_____	_____
c.	_____	_____	_____	_____

51. List key personnel:

	Name	Position	Yr. of Birth	Yrs. Exper.	Previous Employer
a.	_____	_____	_____	_____	_____
b.	_____	_____	_____	_____	_____
c.	_____	_____	_____	_____	_____
d.	_____	_____	_____	_____	_____
e.	_____	_____	_____	_____	_____

52. List any life insurance in effect on key personnel:

	Name	Beneficiary	Amount	Cash Value
a.	_____	_____	$_____	$_____
	Insurance Company: _____			
b.	_____	_____	$_____	$_____
	Insurance Company: _____			
c.	_____	_____	$_____	$_____
	Insurance Company: _____			

53. List other insurance coverage currently in effect:

    Limits in '000's

	BI	PD	Carrier	Expiration Date
a. General liability	$_____	$_____	_____	_____
b. Auto liability	$_____	$_____	_____	_____

(*continued*)

```
    c. Umbrella              $ _____   $ _____   _____   _____
    d. Owner's protection    $ _____   $ _____   _____   _____
54. List any subsidiaries and affiliates of the contracting firm:
            Firm Name                    Ownership                  Type Business
    a. _____   _____   _____
    b. _____   _____   _____
    c. _____   _____   _____
55. Remarks: _____
    _____

                                    Completed by: _____
                                    Title: _____
                                    Date: _____
```

and suppliers. The underwriter usually contacts project owners, subcontractors, material suppliers, and banks for input relative to the reputation of the construction company.

Capacity is the management capability, the organizational depth, the adequacy of plant and equipment, and the collective experience. The surety wants to ensure that the company has sufficient experience and resources, both supervisory talent and technical staff, to match the requirements of the project and has or can obtain the equipment necessary to do the work. Past experience on similar type and size of project is also evaluated.

Capital is the adequacy of the construction company's financial resources in terms of job size and total work program. This includes both assets as well as credit facilities, such as lines of credit or loans. The surety wants to ensure that the construction firm has the financial strength to support the desired work program and has an excellent credit history. Financial statements are analyzed, using the techniques that will be discussed in Chapter 4, to determine the capitalization, profitability, and liquidity of the company. In addition, a schedule of contracts in progress, similar to the one illustrated in Figure 3.17, is usually analyzed.

Sureties typically establish a maximum value of uncompleted work that they will bond. This is known as the *bonding capacity* of the construction company. A rough order-of-magnitude value used by many sureties is to limit a construction firm's bonding capacity to about ten times the firm's working capital, which is the difference between the current assets and current liabilities.[1] So if a construction company has working capital of $2 million, a surety may agree to cover a work program of $20 million.

---

[1]Determination of working capital is discussed in more detail in Chapter 4.

FIGURE 3.17 Schedule of Contracts in Progress

Western States Construction Company
Work in Progress Schedule

Job Description	Contract Price Including Approved Change Orders	Original Estimated Gross Profit	Total Billings to Date Including Retainage	Total Costs to Date	Current Estimate of Costs to Complete	Completion Date (mo/year)
Northside Library	$10,675,000	$600,000	$8,685,780	$8,297,850	$1,736,870	3/2006
North Shore Power Plant	$20,500,000	$1,050,000	$10,967,549	$10,4874,980	$8,387,975	7/2006
Central Street Tower	$12,950,000	$880,000	$5,987,573	$5,597,828	$5,987,376	9/2006
Eastside High School	$11,750,000	$575,000	$2,987,367	$2,864,275	$8,976,482	1/2007

This Form Completed By: *Gabriel T. Nelson*
Date Prepared: January 9, 2006

Summary

Construction is a risky business, and risk management is a significant responsibility of construction company leaders. Business risks include liability for the activities of employees and the potential loss of or damage to company property. One approach is to transfer some of the risk to insurance companies by the purchase of insurance policies or contracts. Surety bonds are also used as risk management tools.

Construction insurance is a highly specialized field, and construction companies rely on insurance agents to help the companies develop comprehensive insurance programs. The primary types of construction insurance can be grouped into four areas: (1) property insurance for the construction project, (2) property insurance for the construction company's property, (3) liability insurance, and (4) business insurance.

Builder's risk insurance provides coverage for financial loss due to damage to the project under construction. Bridge insurance and boiler and machinery insurance may also be needed for certain types of projects. Construction company property insurance includes insurance on company-owned buildings, equipment floater insurance, transportation floater insurance, automobile insurance, and floating marine equipment insurance.

Commercial general liability insurance is purchased to protect against loss due to claims from third parties for injury or property damage. Umbrella insurance is used to provide extra liability insurance coverage in the event of a catastrophic loss. Railroad liability insurance is needed if the construction firm is working on or near railroads. Errors and omissions insurance covers professional negligence of design professionals. Pollution liability insurance provides coverage for legal liability relating to pollution. Employment practices liability insurance and directors and officers liability insurance cover liability resulting from company business practices. Worker's compensation insurance is no-fault insurance that covers financial liability for injury or death of an employee.

Surety bonds are three-party agreements in which the surety guarantees the obligee that the principal will perform its contract obligations. A surety bond is not an insurance policy because the construction company must sign an indemnity agreement to hold the surety harmless for all expenses incurred as a result of the bond being called. Construction companies purchase bonds through bonding agents who have access to multiple surety markets. The commonly used surety bonds in construction are bid bonds, performance bonds, payment bonds, maintenance bonds, and supply bonds. Surety underwriters consider the character, capacity, and capital of a construction firm when deciding the amount of bond coverage to provide.

Review Questions

1. What are the four business risks that a construction company is exposed to?
2. What are the three alternative strategies for managing a risk?
3. How does a construction company transfer some of its risk by purchasing insurance policies?
4. What is the difference between an insurance broker and an insurance company?
5. What is builder's risk insurance, and why is it purchased?
6. Why might the purchaser of a builder's risk insurance policy require that a waiver of subrogation among all insured parties be included in the policy?
7. Why might a construction company need to purchase both equipment floater insurance and automobile insurance?
8. What is commercial liability insurance, and why is it purchased by many construction firms?
9. What is the difference between occurrence-based and claims-based liability insurance policies?
10. What is errors and omissions insurance? Why might a construction company purchase such insurance coverage?
11. What is worker's compensation insurance?
12. How are worker's compensation insurance premiums determined?
13. How does a surety bond differ from an insurance policy?
14. Who are the three parties in a typical surety contract?
15. What is a surety bond indemnity agreement?
16. Why might a construction company purchase surety bonds?
17. What are the responsibilities of a bonding agent?
18. What are the five types of surety bonds used in construction? What are the purposes of each type of bond?
19. What factors are considered by surety underwriters when determining how much bond coverage they will provide to a construction company?

Exercises

1. Make a list of the risks that a highway contractor may face during the construction of a ten-mile section of Interstate highway, which includes the construction of three bridges over water. Select an appropriate risk management strategy for each identified risk.
2. A construction company has a $100 million contract to construct a major industrial plant. What types of insurance policies should the construction company purchase to cover its risks on the project?
3. A construction company has a $300 million contract for the construction of a power plant that will burn natural gas. As a part of its risk management strategy, the construction company has decided to require surety bonds from selected material suppliers and subcontractors. From which suppliers and subcontractors should the construction company require bonds, and what type of bonds should it require?

Sources of Additional Information

Bartholomew, Stuart H. *Construction Contracting: Business and Legal Principles*, 2nd ed., Upper Saddle River, N.J.: Prentice-Hall, 2002.

Clough, Richard H., Glenn A. Sears, and S. Keoki Sears. *Construction Contracting*, 7th ed., Hoboken, N.J.: John Wiley & Sons, Inc., 2005.

Davis, Steven D. and Ron Prichard. *Risk Management, Insurance and Bonding for the Construction Industry*, Alexandria, Va.: Associated General Contractors of America, 2000.

Lewis, Richard C. *Contract Suretyship: From Principle to Practice*, New York: John Wiley & Sons, Inc., 2000.

Palmer, William J., James M. Maloney, and John L. Heffron III. *Construction Insurance, Bonding & Risk Management*, New York: McGraw-Hill, 1996.

Russell, Jeffrey S. *Surety Bonds for Construction Contracts*, Reston, Va.: ASCE Press, 2000.

CHAPTER 4

Financial Analysis and Management

INTRODUCTION

Good financial management skills are essential in the management of a construction company. As mentioned in Chapter 1, the construction industry has the third-highest rate of bankruptcy in the United States. Only the Internet companies and the food service industry have higher bankruptcy rates than does the construction industry. Many construction firms declare bankruptcy not because they are not profitable, but because they do not have sufficient cash to meet their financial obligations. Just as a human cannot survive without an adequate supply of blood, a construction firm cannot survive without an adequate supply of cash. Therefore, cash management is an important aspect of financial management.

Many construction firm managers have strong technical skills, but may have weak financial management skills. This often leads to inadequate attention being placed on the financial condition of the company, until it is too late to take corrective action. The purpose of this chapter is to provide a basic understanding of the principles of financial analysis and management. It is not to tell the reader how to be an accountant or a chief financial officer for a company. It is to provide an overview of the concepts and principles that should be understood by all company leaders. First we will discuss accounting systems and how they are used to collect and organize financial data. Then we will examine financial statements and how they are used to assess the financial condition of the company. Lastly we will discuss cash and capital asset management, operations budgeting, and various sources for obtaining external financing.

ACCOUNTING SYSTEMS

Accounting is the process of collecting, analyzing, classifying, and accumulating historical financial data in categories that will reflect the financial condition of a company's operation. The accounting system is an organizational structure used for the collection and classification of the financial data. The system design should be based on

- the organizational structure of the company,
- the functions that are performed within the company, and
- the type of management reports needed by company managers.

Primary Accounting System

The primary accounting system is known as a **general ledger system**, which is used by an accountant to generate financial statements. The basic accounts within the general ledger system are

- Assets
- Liabilities
- Owners' Equity
- Revenue
- Expenses

Secondary or subsidiary systems are used to generate management reports, such as accounts receivable, accounts payable, or project cost reports.

The general ledger system is based on the following equation:

$$\text{assets} = \text{equities}$$

Assets are things of value, which are owned by the company, and equities are claims to the assets of the company by creditors and owners. The claims of creditors are known as **liabilities**, and those of the owners are known as **owners' equity**. Owners' equity is also known as the net worth or capitalization of the company.

The first step in establishing an accounting system is the creation of a **chart of accounts** for the company general ledger system. This is begun by creating a separate numerical account for each category of financial data that will be a part of the financial statement or management report. Then subaccounts are created for all categories of financial data that are to be tracked and monitored. Subaccounts should be created only for information that is needed to manage the company. Otherwise, resources are wasted in tracking information that is not needed for management decisions. A sample chart of accounts that might be used by a small construction company is shown in Figure 4.1. A chart of accounts is used for the

FIGURE 4.1 Example of a Chart of Accounts for a Construction Company

Chart of Accounts	
Code	**Accounting Category**
100	**Assets**
110	Cash
112	Cash on Hand
114	Cash in Bank
120	Short-Term Investments
130	Receivables
132	Receivables from Current Contracts
134	Retention
136	Receivables from Sale of Vehicles and Equipment
140	Inventory
150	Construction Work in Progress

(*continued*)

Code	Accounting Category
154	Direct Construction Cost
156	Indirect Construction Cost
158	Warranty Cost
160	Other Current Assets
170	Investments
180	Property, Plant, and Equipment
182	Buildings and Land
184	Construction Equipment
186	Administrative Vehicles
188	Office Equipment
190	Accumulated Depreciation
200	**Liabilities and Owners' Equity**
210	Accounts Payable
212	Accounts Payable to Subcontractors
214	Accounts Payable to Suppliers
216	Accounts Payable on Vehicles and Equipment
220	Notes Payable
230	Other Current Liabilities
240	Long-Term Liabilities Payable
250	Owners' Equity
300	**Revenues and Cost of Sales**
310	Revenues
320	Cost of Sales
400	**Financing Expense**
500	**Marketing Expense**
600	**General and Administrative Expense**
610	Salaries
620	Office Expense
630	Computer Expense
640	Professional Services
650	Insurance
660	Taxes
662	Payroll Taxes
664	FICA Taxes
666	Income Taxes
668	Property Taxes
670	Depreciation Expense
700	**Other**

creation of all company accounting records. A separate accounting record is maintained for each code listed in the chart of accounts. Each accounting record will show income and expenses for each code or account.

Secondary Accounting Systems

One type of secondary accounting system is a project cost ledger system, which is used to account for financial data relative to individual construction projects. Both project costs and revenues are tracked in this system to allow company managers to monitor

CHAPTER 4 Financial Analysis and Management

Project Cost Control Chart of Accounts	
Code	**Desciption**
01100	Project Supervision
01200	Temporary Facilities
01300	Temporary Utilities
01400	Project Support Equipment
02100	Site-Work Demolition
02200	Earthwork
02300	Utility Line Construction
02400	Paving and Surfacing
02500	Landscaping
03100	Cast-in-Place Concrete
03200	Precast Concrete
03300	Tilt-up Concrete Panels
04100	Brick Masonry
04200	Concrete Unit Masonry
04300	Stonework
05100	Structural Steel
05200	Metal Joists
05300	Metal Decking
05400	Miscellaneous Metal
06100	Rough Carpentry
06200	Finish Carpentry
06300	Casework
06400	Stair Work
07100	Waterproofing
07200	Insulation
07300	Fire Proofing
07400	Roofing
08100	Metal Doors and Frames
08200	Wood Doors
08300	Windows
09100	Drywall
09200	Tile
09300	Floor Covering
09400	Painting
10100	Partitions
10200	Toilet and Bath Accessories
11000	Appliances
14000	Elevators
15000	Mechanical
16000	Electrical

FIGURE 4.2 Example of a Chart of Accounts for a Project Cost Ledger

the financial status of individual projects. A typical chart of accounts for a project cost ledger system is shown in Figure 4.2. Another secondary accounting system used by many construction companies is an equipment cost ledger, which is used to determine the hourly rates to be charged to individual projects for the use of company-owned equipment. Equipment cost ledgers are used to keep track of all costs associated with each item of owned equipment.

An example of a project cost coding system, using the chart of accounts shown in Figure 4.2, is

$$\underline{\text{(project number)}} . \underline{\text{(chart of accounts code)}} . \underline{\text{(element of cost)}}$$

Here the project number is assigned by the construction firm, the chart of accounts code comes from Figure 4.2, and the element of cost is the type of cost. An example of an element of cost coding is

1. labor
2. equipment
3. incorporated materials
4. subcontract
5. supplies and materials not incorporated into project

Using this system, the cost code for project 9562 for cast-in-place concrete labor would be

9562.03100.1

and for incorporated materials, would be

9562.03100.3

Thus the job cost ledger is structured to track costs and revenue based on the company's work-breakdown structure for each project.

Other types of secondary accounting systems are equipment ledgers, accounts receivable, and accounts payable. These systems are used to generate reports that are used to determine the book value of owned equipment and to track equipment operating costs as well as to monitor accounts receivable and accounts payable.

Accounting Records

Accounting records may be kept in paper form using journals and ledgers, or more typically they are kept electronically in a database. Source documents used by accounting technicians and bookkeepers to input financial data are payroll data, invoices, cash receipts, and requests for progress payments. Paper or electronic journals are maintained with the data received to form the basis for the development of management reports. Typical reports are payroll, accounts receivable, accounts payable, cash receipts, cash disbursements, and job cost reports. A job cost journal accumulates actual project cost data by cost code and element of expense, such as labor, material, equipment, subcontractor, or project overhead. Codes may be further divided into the specific construction task, such as site work or rough carpentry. Collected project cost data can then be used to update the cost-estimating database, which can be used to support the development of future estimates. Journal entries are also posted to the general ledger for the preparation of financial statements.

ACCOUNTING METHODS

A construction company may use either a cash or an accrual method for recording revenues and expenses. The cash method is easier to use, but the accrual method usually provides more accurate information. A **cash method of accounting** reports revenues and expenditures in the accounting period in which cash is received or disbursed, regardless of when the revenues were earned or the expenses incurred. The bookkeeper records only cash transactions, that is cash deposits made to a bank and checks written. Because of the lag between the earning of the revenues and the receipt of cash and between the commitment to an expenditure and the actual payment, management reports and financial statements do not reflect actual financial conditions.

An **accrual method of accounting** recognizes revenues when they are earned and expenses when they are incurred, regardless of when the cash transaction takes place. For example, income would be credited when the request for progress payment was sent, not when the payment was received. Likewise, expenditures would be debited when materials are ordered, not when payments are made. Therefore, the accrual system provides a more accurate picture of the financial condition of the company, and is usually the preferred method of accounting.

There are two methods of construction accounting that a construction firm may use to account for profit on a construction project. One is the percentage-of-completion method, and the other is the completed-contract method. In the **percentage-of-completion method**, the company estimates the percentage completion of each construction project at the end of the accounting period, using the following equation:

$$\frac{\text{cost incurred to date}}{\text{cost incurred to date } + \text{ estimated cost to complete}}$$

This percentage is multiplied by the contract value for the project to determine the revenue generated to date. The total project expenditures to date are then subtracted from the calculated revenue to determine the profit realized during the accounting period. In the **completed-contract method**, the construction company does not take credit for any profit until the project is completed and the contract closed out.

For example, suppose a contractor has a $5 million lump-sum contract for the construction of a high school, and the contract states that the project owner will retain 5 percent of each progress payment until the project is completed. At the end of the first year of construction, the project is estimated to be 50 percent complete, and the contractor has earned $2,375,000, which is 50 percent of the contract value less 5 percent retainage. The contractor's total project cost during the first year was $2,350,000. In the percentage-of-completion method, the contractor would report a net profit before taxes of $25,000 for the project, which is $2,375,000 less $2,350,000. In the completed-contract method, the contractor would report no profit until the project was closed out, and the actual profit had been determined. The percentage-of-completion method provides a more accurate assessment of the company's financial condition and is used by most construction firms in developing financial statements for lending institutions and bonding companies. The completed-contract method is often preferred when preparing financial statements for companies with numerous short-term contracts, such as subcontractors, because profits are not accurately known until the contracts are closed out. The completed-contract method is usually not used by general contractors.

FINANCIAL STATEMENTS

Financial statements provide a picture of the financial condition of the construction company. They are used by managers, bankers, bonding companies, and customers in assessing the financial health of the firm. The ability to read and understand these statements is an essential business management skill. The two most commonly used financial statements are the balance sheet and the income statement.

Balance Sheet

The **balance sheet** represents the financial condition of the company as of the date of the balance sheet, usually the end of a month, a quarter, or a year. A typical balance sheet is illustrated in Figure 4.3. This balance sheet was developed for an

FIGURE 4.3 Balance Sheet for Western States Construction Company

	Current Year	Last Year
ASSETS		
Current Assets		
Cash	$809,675	$870,650
Accounts Receivable—Trade	$1,863,150	$1,287,690
Accounts Receivable—Retention	$172,200	$105,875
Inventory	$35,785	$23,487
Prepaid Expenses	$20,769	$25,345
Total Current Assets	$2,901,579	$2,313,047
Fixed Assets		
Land	$165,890	$150,735
Buildings	$785,970	$697,658
Furniture and Equipment	$150,987	$140,685
Motor Vehicles	$224,775	$201,425
Construction Equipment	$980,725	$875,943
Subtotal	$2,308,347	$2,066,446
Less Accumulated Depreciation	($925,340)	($797,780)
Total Fixed Assets	$1,383,007	$1,268,666
TOTAL ASSETS	$4,284,586	$3,581,713

(continued)

FIGURE 4.3 (continued)

	Current Year	Last Year
LIABILITIES AND OWNERS' EQUITY		
Current Liabilities		
Notes Payable	$100,000	$75,000
Accounts Payable—Trade	$967,050	$894,300
Accounts Payable—Retention	$35,600	$88,400
Accrued Payables	$114,350	$144,775
Accrued Taxes	$67,950	$60,125
Current Maturity—Long-Term Debt	$125,000	$95,000
Total Current Liabilities	$1,409,950	$1,357,600
Long-Term Debt	$450,000	$500,000
TOTAL LIABILITIES	$1,859,950	$1,857,600
OWNERS' EQUITY		
Capital Stock	$500,000	$500,000
Retained Earnings	$1,924,636	$1,224,113
TOTAL OWNERS' EQUITY	$2,424,636	$1,724,113
TOTAL LIABILITIES AND OWNERS' EQUITY	$4,284,586	$3,581,713

accrual accounting system using the percentage-of-completion accounting method. By definition, accounts shown on the balance sheet must balance in accordance with the following equation:

$$\text{assets} = \text{liabilities} + \text{owners' equity}$$

Assets are resources owned by the company that may be converted in the future to cash inflows. Liabilities are obligations for future cash outflows. Owners' equity is the net worth of the company after all liabilities are paid and is calculated by subtracting the company's liabilities from its assets. Owners' equity includes both the owners' investment in the company and company earnings that have not been paid to the owners (retained earnings). Owners' equity is also known as the capitalization of the company.

    Current assets are non-depreciable assets, such as cash, accounts receivable, and inventory. Cash would include demand and time deposits in financial institutions (checking and savings accounts) and cash on hand. **Accounts receivable** would include claims against others for work performed or assets sold, including requests for progress payment not paid and retainage. Inventory would include materials purchased but not yet used, either for construction operations or to meet other company requirements.

Fixed assets are depreciable assets that will be retained for longer than one year. This may include vehicles, construction equipment, buildings and land, as well as office equipment. The value shown on the balance sheet is the net **book value** of depreciable assets after accumulated depreciation has been subtracted. Depreciation is a portion of the acquisition cost of an asset, such as equipment or a building, that can be written each year due to age and use. Other assets may include long-term investments.

Current liabilities are debts that are expected to be paid within a year. **Accounts payable** are debts owed to suppliers and subcontractors or for the purchase of any good or service. Notes payable are obligations to pay for money borrowed to meet operating requirements, such as from a line of credit. Accrued payables include rent due on an office, taxes due, or salaries payable. Long-term liabilities are long-term debt, such as mortgages or equipment loans.

Income Statement

The other type of financial statement is an **income statement**, such as the one illustrated in Figure 4.4. It summarizes the profitability of the company over a period of time, listing all sources and amounts of revenue and subtracting the cost of sales and operating expenses to arrive at the net income. This income statement was developed for an accrual accounting system using the percentage-of-completion accounting method.

The following equations summarize the data shown on an income statement:

1. **gross profit** = sales − cost of sales
2. **net profit before taxes** = gross profit − operating expenses
3. **net profit after taxes** = net profit before taxes − taxes

FIGURE 4.4 Income Statement for Western States Construction Company

	Current Year	Last Year
Revenues		
Sales	$22,698,572	$18,875,342
Cost of Sales	($17,854,395)	($15,150,360)
Gross Profit	$4,844,177	$3,724,982
Operating Expenses		
Financing	$250,875	$150,230
Marketing	$376,980	$250,760
General and Administrative	$2,796,845	$2,167,549
Operating Expenses	$3,424,700	$2,568,539
Net Profit before Taxes	$1,419,477	$1,156,443
State and Federal Taxes	($236,579)	($192,740)
Net Profit after Taxes	$1,182,898	$963,703

Revenue is the income earned during the period based on the portion of each contract completed, whether or not payment has been received. Cost of sales includes all direct and indirect project costs incurred, whether or not payment has been made. Project direct costs include materials, labor, equipment, and subcontractor costs. **Project indirect costs** include project costs that are not allocated to a specific element of work, such as project overhead (job supervision, project office, bonding, etc.).

Please note that the balance sheet is prepared to reflect the financial condition of the company on a specific date, while the income statement covers a specific time period.

FINANCIAL ANALYSIS

There are two major aspects of analyzing the financial condition of a company. First is the analysis of the financial statements to assess the overall financial health of the company. The other is a cash flow analysis to determine cash flow requirements, so financial management strategies can be crafted to ensure that the firm has adequate cash resources to meet anticipated requirements.

Financial Statement Analysis

The major financial characteristics of interest in evaluating financial statements are liquidity, profitability, and leverage or debt-to-equity. **Ratio analysis** is used to assess these characteristics at single points in time, and trend analysis is used to determine how they have changed over time. In other words, has profitability increased, decreased, or remained the same over the past five years? Typical ratios for various industries are published annually by several sources. Dun & Bradstreet publish one such source, *Industry Norms and Key Business Ratios*, which was used to create the typical value figures provided in this section. Data from 1997 through 2005 were analyzed to create the values shown.

The gross sales, the cost of sales, the operating expenses, and the net profit for each accounting period can be determined from the income statement. These should be analyzed over time to determine trends. If profitability is declining while sales are increasing, it may mean that the price the company charges for its services may be too low, the cost of performing construction was estimated too low, the lower productivity is resulting in increased construction costs, or the company overhead costs have grown faster than revenues.

Liquidity Analysis

Liquidity measures the ability of a company to meet its current obligations when they become due. **Liquidity analysis** is an assessment of the liquidity condition of a company. There are two ratios that are typically determined to assess the liquidity of the company—the current ratio and the quick ratio. The **current ratio** is most commonly used to measure liquidity. It is defined as

$$\text{current ratio} = \frac{\text{current assets}}{\text{current liabilities}}$$

The current ratio measures the company's ability to pay current liabilities and have enough working capital to continue operations. **Working capital** is the liquid assets that the company has available to finance its future operations. It is defined as

$$\text{working capital} = \text{current assets} - \text{current liabilities}$$

A company's working capital should be at least 5 to 10 percent of the company's annual revenues. Typical current ratios for construction firms are shown in Figure 4.5. The top 25 percent means that 25 percent of the companies surveyed had ratios equal to or better than the value shown, the median value means that 50 percent of the companies surveyed had ratios equal to or better than the value shown, and the lower 25 percent means that 25 percent of the companies surveyed had ratios equal to or worse than the value shown.

Let's look at the current ratio for Western States Construction Company for the current year. Looking at the balance sheet shown in Figure 4.3, we find that the current assets are $2,901,579 and the current liabilities are $1,409,950. The current ratio is then found to be

$$\frac{\$2,901,579}{\$1,409,950} = 2.06$$

which indicates that the company has good liquidity (above the median value for companies surveyed) when compared to the values shown in Figure 4.5. The current year working capital is

$$\$2,901,579 - \$1,409,950 = \$1,491,629$$

From the income statement in Figure 4.4, we determine the company's annual revenue to be $22,698,572. Checking to see if the company has adequate working capital, we divide the working capital by the total revenue, as follows:

$$\frac{\$1,491,629}{\$22,698,572} = 0.066$$

We determine that the working capital is 6.6 percent of the annual revenue (which is between 5 percent and 10 percent), so we conclude that the company has adequate working capital.

The **quick ratio** is defined as

$$\text{quick ratio} = \frac{\text{current assets} - \text{inventory}}{\text{current liabilities}}$$

FIGURE 4.5 Typical Current Ratios for Construction Firms

Construction Sector	Top 25%	Median	Lower 25%
Residential	2.5	1.4	1.1
Commercial	2.2	1.5	1.2
Heavy Civil	2.7	1.7	1.3
Specialty Construction	3.2	1.9	1.4

FIGURE 4.6 Typical Quick Ratios for Construction Firms

Construction Sector	Top 25%	Median	Lower 25%
Residential	1.6	0.8	0.2
Commercial	1.6	1.1	0.6
Heavy Civil	1.9	1.2	0.7
Specialty Construction	2.3	1.4	0.8

It is a measure of the company's ability to extinguish its current liabilities quickly through the use of current assets that can be quickly turned into cash. The quick ratio is often called the acid test because it considers the liquidity of the components of the current assets. Typical quick ratios for construction companies are shown in Figure 4.6.

Let's look at the quick ratio for the current year for Western States Construction Company. From Figure 4.3, we find that the inventory is $35,785, so the quick ratio is calculated to be

$$\frac{\$2,901,579 - \$35,785}{\$1,409,950} = 2.03$$

which is not much different than the current ratio, because the company has little inventory. When compared to the values shown in Figure 4.6, the quick ratio for Western States Construction Company is quite good (better than the top 25 percent value of 1.6 for commercial construction firms).

Profitability Analysis
Profitability analysis is an analysis of the ability of a company to generate profits from its operations. The **return on assets** is a measure of the productivity of the company. It is determined by the following ratio:

$$\text{return on assets} = \frac{\text{net profit after taxes}}{\text{total assets}}$$

It is a measure of the overall efficiency of the company in managing assets and generating profits. Typical returns on assets for construction firms are shown in Figure 4.7.

FIGURE 4.7 Typical Returns on Assets for Construction Firms

Construction Sector	Top 25%	Median	Lower 25%
Residential	20.3%	7.3%	1.6%
Commercial	11.4%	4.1%	0.9%
Heavy Civil	11.5%	4.5%	0.6%
Specialty Construction	16.7%	6.0%	1.1%

Let's look at the return on assets for the current year for the Western States Construction Company. From Figure 4.4, we find the net profit to be $1,182,898, and from Figure 4.3, we find the total assets to be $4,284,586. Thus the return on assets is found to be

$$\frac{\$1,182,898}{\$4,284,586} = 0.28 \text{ or } 28\%$$

which is an excellent return when compared to the values shown in Figure 4.7.

The **return on equity** is a measure of the return on the owners' investment. It is determined by the following ratio:

$$\text{return on equity} = \frac{\text{net profit after taxes}}{\text{owners' equity}}$$

Typical returns on equity for construction firms are shown in Figure 4.8.

Let's look at the return on equity for the current year for the Western States Construction Company. From Figure 4.3, we find the total equity or net worth to be $2,424,636. The return on equity is then found to be

$$\frac{\$1,182,898}{\$2,424,636} = 0.48 \text{ or } 48\%$$

which again is an excellent return when compared to the data shown in Figure 4.8.

The **return on sales** provides the net profit margin for the company. It is determined by the following ratio:

$$\text{return on sales} = \frac{\text{net profit after taxes}}{\text{sales}}$$

It measures the profit generated after consideration of all expenses. Typical returns on sales for construction firms are shown in Figure 4.9.

Let's look at the return on sales for the current year for the Western States Construction Company. From, Figure 4.4, we find the total sales to be $22,698,572, and the return on sales is

$$\frac{\$1,182,898}{\$22,698,572} = 0.052 \text{ or } 5.2\%$$

which is an excellent return on sales when compared to the values shown in Figure 4.9.

FIGURE 4.8 Typical Returns on Equity for Construction Firms

Construction Sector	Top 25%	Median	Lower 25%
Residential	58.4%	22.3%	6.0%
Commercial	31.5%	11.6%	2.6%
Heavy Civil	26.5%	11.5%	2.1%
Specialty Construction	36.5%	13.1%	2.9%

FIGURE 4.9 Typical Returns on Sales for Construction Firms

Construction Sector	Top 25%	Median	Lower 25%
Residential	5.6%	2.4%	0.5%
Commercial	3.5%	1.3%	0.2%
Heavy Civil	5.8%	2.1%	0.3%
Specialty Construction	5.7%	2.1%	0.4%

Debt Analysis

Debt analysis is an analysis of the amount of debt a company is using to finance its operations. The use of borrowed funds, which is known as **leverage**, may allow a construction company to earn a higher return on owners' equity than would be possible without the use of the borrowed capital. This is true only if the company can earn a greater return on the borrowed funds than what it must pay as interest for the use of them. Too much debt is risky, particularly if interest rates are high.

The **debt-to-equity ratio** is a measure of the leverage of the company. It is determined by the following ratio:

$$\text{debt-to-equity ratio} = \frac{\text{total liabilities}}{\text{owners' equity}}$$

The debt-to-equity ratio compares the amount of capital provided by creditors to the equity or capitalization of the firm. A high ratio indicates a heavy dependency on debt, with the associated risks to creditors. Typical debt-to-equity ratios for construction firms are shown in Figure 4.10.

Let's look at the debt-to-equity ratio for the current year for the Western States Construction Company. From Figure 4.3, we find that the total liabilities are $1,859,950 and the total equity is $2,424,636. Thus, the debt-to-equity ratio for the company is

$$\frac{\$1,859,950}{\$2,424,636} = 0.77 \text{ or } 77\%$$

which is an excellent debt-to-equity ratio when compared to the values shown in Figure 4.10. This ratio is very important to sureties when determining the amount of bonding capacity to provide to a construction firm. The ratio should be less than 2 to demonstrate that the owners have more in the company than do the creditors.

FIGURE 4.10 Typical Debt-to-Equity Ratios for Construction Firms

Construction Sector	Top 25%	Median	Lower 25%
Residential	63%	164%	385%
Commercial	73%	155%	304%
Heavy Civil	46%	102%	202%
Specialty Construction	43%	97%	199%

Trend Analysis

Trend analysis is examining the financial performance of the company over time. A financial trend is an upward or a downward movement of certain financial data in successive time periods, usually a year. The objective is to identify positive or negative developments that affect the company's financial strength. Trend analysis involves the following steps:

- Select the data to track
- Collect the data
- Analyze the data over time

Now let's look at an example. The leaders of Western States Construction Company decided to collect the data shown in Figure 4.11 to serve as a basis for their analysis. The data can be depicted in tabular form, as shown in the figure, or in graphical form.

Analysis of the data in Figure 4.11 indicates that the company had a 71 percent growth in sales over the five-year period with a resulting 157 percent growth in net profits after taxes. Working capital grew by 144 percent, and owners' equity increased by 145 percent. Company liquidity also improved as the current ratio increased from 1.5 to 2.06. The primary reasons for this improved financial performance were the increased return on sales and the reduction in the debt-to-equity ratio. Overall, the financial condition of the company significantly improved over the five-year period.

Cash Flow Analysis

A construction company is totally dependent on maintaining a sufficient level of cash to finance its daily operations. If adequate cash reserves are not maintained, the company must either borrow from lending institutions or default on its obligations. If

FIGURE 4.11 Selective Financial Data for Western States Construction Company

Financial Measure	Current Year-4	Current Year-3	Current Year-2	Current Year-1	Current Year
Total Sales	$13,271,608	$15,275,362	$17,136,547	$18,875,342	$22,698,572
Cost of Sales	$11,736,205	$13,165,268	$14,276,302	$15,150,360	$17,854,395
Operating Expenses	$976,321	$1,536,275	$1,896,275	$2,568,539	$3,424,700
Net Profit after Taxes	$459,715	$477,991	$801,987	$963,703	$1,182,898
Working Capital	$612,067	$696,371	$853,167	$955,447	$1,491,629
Owners' Equity	$988,675	$996,486	$1,637,894	$1,724,113	$2,424,636
Current Ratio	1.50	1.55	1.65	1.70	2.06
Quick Ratio	1.49	1.53	1.64	1.67	2.03
Return on Assets	16%	15%	23%	27%	28%
Return on Equity	46%	48%	49%	56%	49%
Return on Sales	3.5%	3.1%	4.7%	5.1%	5.2%
Debt-to-Equity Ratio	173%	177%	111%	109%	77%

the cost of borrowing was not included in a project budget, the interest charges on borrowed cash may significantly erode the anticipated profit. The cash flow cycle within the company is illustrated in Figure 4.12. Cash is invested in projects in progress, which then become accounts receivable. The accounts receivable eventually are collected and become cash. Some portion of the accounts receivable may be retained by the project owners until the end of the projects, even though the construction firm made cash disbursements to cover project costs.

Because of the lag in receiving payment for accounts receivable (usually 30 to 60 days), the net cash flows on projects tend to be negative during the early phases of a project and positive during the later phases. Predicting cash flow requirements involves conducting a **cash flow analysis** for each project. Project expenditures and receivables are estimated for each month, and the difference between expenses and receivables determines the net cash flow requirements, as illustrated in Figure 4.13

FIGURE 4.12 Cash Flow Cycle

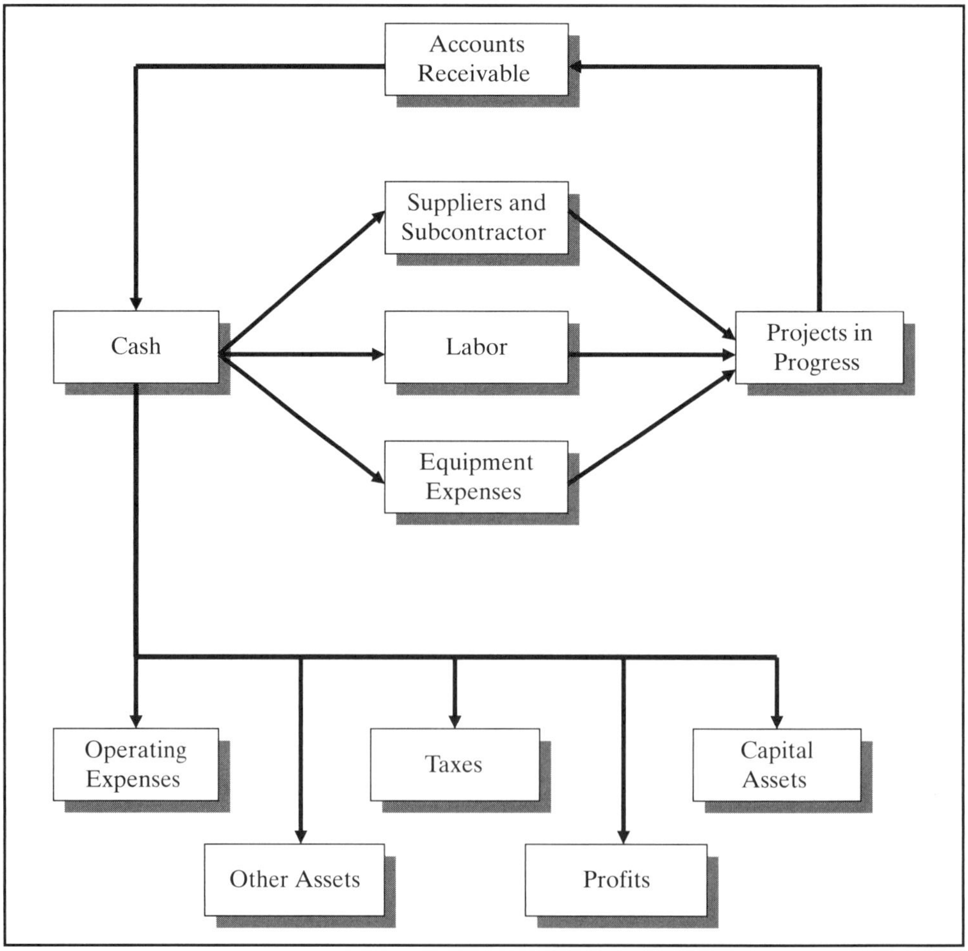

FIGURE 4.13 Cash Flow Analysis for Job #1

	Estimate	January	February	March	April	May	June	July	August	September	October	November
Total Job Revenue												
Total Draw	$1,000,000	$60,000	$80,000	$100,000	$120,000	$140,000	$150,000	$120,000	$100,000	$80,000	$50,000	
Retainage	$50,000	$(3,000)	$(4,000)	$(5,000)	$(6,000)	$(7,000)	$(7,500)	$(6,000)	$(5,000)	$(4,000)	$(2,500)	
Net Receipt	$1,000,000		$57,000	$76,000	$95,000	$114,000	$133,000	$142,500	$114,000	$95,000	$76,000	$97,500
Direct Job Costs												
Labor	$25,500	$1,530	$2,040	$2,550	$3,060	$3,570	$3,825	$3,060	$2,550	$2,040	$1,275	
Material	$176,700	$10,602	$14,136	$17,670	$21,204	$24,738	$26,505	$21,204	$17,670	$14,136	$8,835	
Subcontractor	$632,000	$37,920	$50,560	$63,200	$75,840	$88,480	$94,800	$75,840	$63,200	$50,560	$31,600	
Equipment	$10,700	$642	$856	$1,070	$1,284	$1,498	$1,605	$1,284	$1,070	$856	$535	
Total Direct Job Costs	$844,900	$50,694	$67,592	$84,490	$101,388	$118,286	$126,735	$101,388	$84,490	$67,592	$42,245	
Net Job Cash Flow		$(50,694)	$(10,592)	$(8,490)	$(6,388)	$(4,286)	$6,265	$41,112	$29,510	$27,408	$33,755	$97,500
Cumulative Cash Flow	$155,100	$(50,694)	$(61,286)	$(69,776)	$(76,164)	$(80,450)	$(74,185)	$(33,073)	$(3,563)	$23,845	$57,600	$155,100

83

CHAPTER 4 Financial Analysis and Management

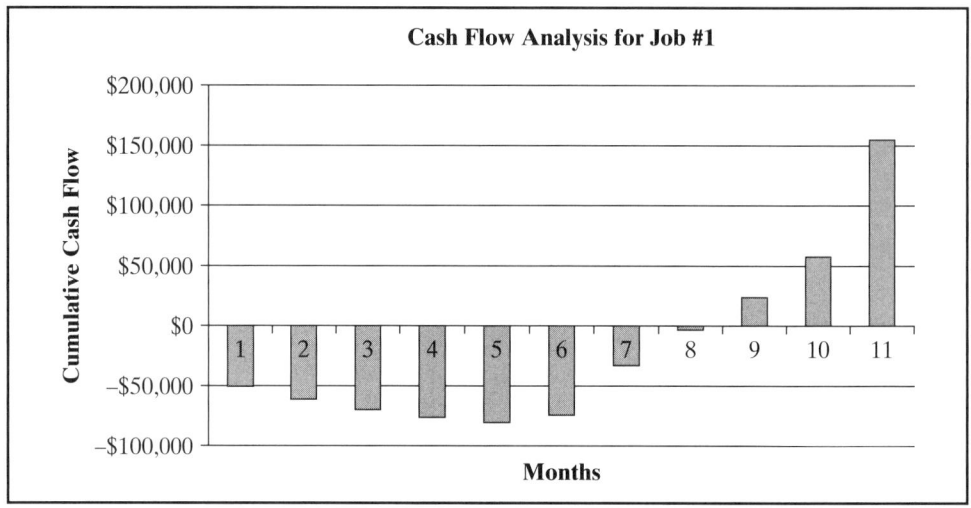

FIGURE 4.14 Cumulative Cash Flow for Job #1

for Job #1. The project has a contract value of $1 million and a ten-month duration. The construction contract stipulates a 5 percent retention on each progress payment. A plot of the cumulative cash flow requirements for the project is shown in Figure 4.14. Note that cash flow for Job #1 is negative during the early months of the project and positive during the later months.

The monthly cash flow requirements for each project are summed in a spreadsheet, as shown in Figure 4.15, to determine overall cash requirements for the company. Negative cash flow requirements (those shown in parentheses) must be met with non-project financing, either company reserves or external financing. The most frequently used vehicle for external financing is a line of credit, which is described in the next section.

FINANCIAL MANAGEMENT

A key component of financial management is the development of an annual operations budget, which is used to forecast anticipated revenues and expenses. An operations budget is a numerical plan of operations that is used as a benchmark in evaluating the company's financial performance during the year. The major components of an operating budget are

- Revenue
- Project Costs
- Company Overhead
- Profits

FIGURE 4.15 Company Cash Flow Analysis

Projects	January	February	March	April	May	June	July	August	September	October
Beginning Balance	$200,000	$186,306	$22,714	$(95,776)	$(137,164)	$(125,950)	$(163,185)	$(134,073)	$(127,563)	$(80,855)
In Process:										
Job #1	$(50,694)	$(10,592)	$(8,490)	$(6,388)	$(4,286)	$6,265	$41,112	$29,510	$27,408	$33,755
Job #2	$120,000	$—	$—	$—	$—	$—	$—	$—	$—	$—
Job #3	$(48,000)	$(22,000)	$(15,000)	$20,000	$45,000	$1,500	$65,000	$—	$—	$—
Job #4	$—	$(40,000)	$(25,000)	$(10,000)	$5,000	$15,000	$26,000	$30,000	$35,000	$50,000
Job #5	$—	$(56,000)	$(35,000)	$(10,000)	$500	$25,000	$35,000	$26,000	$35,000	$28,000
Projected New Jobs:										
Job #6	$—	$—	$—	$—	$—	$(50,000)	$(35,000)	$(20,000)	$(5,000)	$6,000
Job #7	$—	$—	$—	$—	$—	$—	$(40,000)	$(15,000)	$(10,000)	$4,000
Job #8	$—	$—	$—	$—	$—	$—	$(28,000)	$(9,000)	$(700)	$5,000
Net Cash Flow	$21,306	$(128,592)	$(83,490)	$(6,388)	$46,214	$(2,235)	$64,112	$41,510	$81,708	$126,755
Total Cash Available	$221,306	$57,714	$(60,776)	$(102,164)	$(90,950)	$(128,185)	$(99,073)	$(92,563)	$(45,855)	$45,900
Operating Expenses	$35,000	$35,000	$35,000	$35,000	$35,000	$35,000	$35,000	$35,000	$35,000	$35,000
Ending Cash Balance	$186,306	$22,714	$(95,776)	$(137,164)	$(125,950)	$(163,185)	$(134,073)	$(127,563)	$(80,855)	$10,900

Actual performance should be measured each month and compared to budgeted values as illustrated in Figure 4.16. This budget was developed using an accrual accounting system and a percentage-of-completion accounting method. Note that while company revenues to date are only 5 percent less than the value anticipated in the budget, the net profit before taxes is 63 percent below the budgeted amount.

Construction companies grow by increasing their capitalization or owners' equity. This can be accomplished either by an infusion of capital, such as from the sale of additional stock or from additional investment, or by retaining a portion of the company's profits, which are called **retained earnings**. Earnings that are retained are not available for disbursement to owners or stockholders. Selection of an appropriate strategy for financing company growth is an important company management issue. The other major issues in financial management are capital asset acquisition, cash management, overhead budgeting, and debt management.

Capital assets are depreciable assets, such as vehicles, equipment, or office facilities. A portion of the value of these assets is charged to company revenues as a depreciation cost. The amount of depreciation depends on the depreciation technique used, the value of the asset, the depreciation period, and the anticipated salvage value, if any. Purchasing capital assets for cash is generally not a good strategy, as it may deplete the company's working capital such that it is unable to finance its cash flow requirements. Working capital is the excess of current assets over current liabilities. A better strategy is to finance the acquisition of the capital assets and

FIGURE 4.16 Operations Budget for Western States Construction Company

	Annual Budget	**Year-to-Date Budget**	**Year-to-Date Actual**	**Variance**
Revenue	$23,000,000	$6,900,000	$6,548,600	$351,400
Direct Cost	$18,000,000	$5,400,000	$5,297,500	$102,500
Gross Profit	$5,000,000	$1,500,000	$1,251,100	$248,900
Financing Expenses	$275,000	$82,500	$75,978	$6,522
Marketing Expenses	$400,000	$120,000	$122,759	−$2,759
General & Administrative Expenses				
Salaries	$650,000	$195,000	$206,865	−$11,865
Travel	$150,000	$45,000	$37,800	$7,200
Office Expenses	$400,000	$120,000	$118,765	$1,235
Computer Expenses	$150,000	$45,000	$41,798	$3,202
Insurance	$600,000	$180,000	$186,256	−$6,256
Professional Services	$150,000	$45,000	$37,865	$7,135
Taxes	$180,000	$54,000	$53,875	$125
Depreciation	$450,000	$135,000	$137,854	−$2,854
Administrative Vehicles	$268,000	$80,400	$76,567	$3,833
Warranty Expenses	$60,000	$18,000	$13,876	$4,124
Total	$3,733,000	$1,119,900	$1,110,258	$9,642
Net Profit before Taxes	$1,267,000	$380,100	$140,842	$239,258

charge their use (vehicles and equipment) to projects. The annual earnings of the equipment and vehicles should be used to pay the annual financing costs. If the earning ability of the vehicle or item of equipment is not projected to exceed the annual financing cost, the item should be leased or rented and not purchased. Project charges for the annual cost of owning office facilities are included in the overhead rate established by the company.

Cash management was discussed in the preceding section. Adequate cash resources are needed to finance the company operations. As a contingency to cover either unanticipated cash flow requirements or projected cash requirements, most construction firms obtain a line of credit from a bank. The line of credit is similar to a credit card. The construction firm pays a fee, say $1,000, to the bank for the right to use a sum of money, say $500,000, for a year. Interest costs on the use of the money are not incurred until actually used. The construction firm uses the line of credit to cover cash flow requirements that cannot be met with internal resources, and then repays the bank as income is received. Interest charges are due only for the time period the money is actually used, similar to a personal credit card.

Another major financial management issue is the development of a budget for the overhead operations of the company. This budget would include the salaries of personnel who are not assigned to specific projects, such as the president, the marketing department, the accounting department, and human resources management, and the costs of operating the home office of the company. An example of an overhead budget is shown in Figure 4.17.

The overhead rate charged to projects is determined by dividing the annual overhead budget plus profit by total anticipated direct project costs. For example,

Element of Expense	Annual Budget
Charitable Contributions	$15,000
Computer Equipment	$45,000
Custodial Services	$22,500
Employee Recruiting	$22,000
Employee Taxes	$75,000
Employee Training	$45,000
Financing Expenses	$295,000
Furniture Expenses	$24,000
Insurance	$125,000
Legal Services	$23,000
License Fees	$2,000
Marketing Expenses	$320,000
Office Rent	$50,000
Office Supplies	$25,000
Office Utilites	$10,500
Postage	$5,000
Salaries	$865,000
Telephone	$10,500
Travel Expenses	$20,500
Total	$2,000,000

FIGURE 4.17 Example of an Overhead Budget

suppose the annual overhead budget is $2,000,000, the desired profit is $500,000, and the total anticipated direct costs for projects is $10,000,000. The overhead rate to be charged to projects would then be

$$\frac{\$2,000,000 + \$500,000}{\$10,000,000} = 0.25 \text{ or } 25\%$$

This means that the overhead markup rate used on all construction project estimates would be 25 percent. This may be too high and make the contractor less competitive in the market. If this rate is used, but actual overhead costs exceed the budget, company profits will decrease. The overhead budget is established at the beginning of the year based on what company leaders believe is affordable, considering the anticipated project workload. If the demand for construction services is anticipated to decline, the company must reduce its overhead budget to remain both competitive and profitable.

Company debt generally includes either transaction loans used to purchase specific item, such as vehicles or equipment, or traditional loans. Transaction loans may be procured from a bank, from the seller, or from a financing company, and the purchased items act as collateral for the loans. Traditional loans generally are obtained from banks without collateral. Care must be exercised to ensure that the amount of traditional loans is minimized, because the cost of servicing these loans will deplete cash reserves.

Summary

To be successful, construction company managers must understand and be able to apply the principles of financial analysis and management. Accounting systems are used to collect and organize financial data. A chart of accounts is used to identify all financial accounts that are to be tracked and monitored. A general ledger system is used to record financial transactions and to serve as a basis for the preparation of financial statements. Either a cash or an accrual method may be used for recording revenues and expenses. The cash method is easier to use, but the accrual method provides more accurate data. Construction companies may use either the percentage-of-completion method or the completed-contract method for accounting for profits on construction projects, but the percentage-of-completion method is usually used because it provides a better assessment of the financial condition of the company.

Financial statements provide a picture of the financial condition of the company on a specific date. The balance sheet and the income statement are the most commonly used financial statements. These financial statements are analyzed to determine profitability, liquidity, return on investment, and leverage. Ratio analysis is used to determine the current financial condition, and trend analysis is used to assess how the firm's financial condition has changed over time. Cash flow analysis is used to determine net cash flow requirements and to select appropriate means for covering negative cash flow requirements.

Financial management strategies are needed for increasing the company's net worth, for purchasing capital assets, for developing operations budgets, for controlling company overhead costs, and for obtaining external financing.

Review Questions

1. What is the basic purpose of an accounting system?
2. What are the five basic accounts used in creating a general ledger system?
3. What criteria should a construction company use in establishing a chart of accounts?
4. Financial data are recorded in both journals and the general ledger. What is the purpose of a journal?
5. How would you create a chart of accounts for a project cost ledger?
6. What is the difference between the cash and the accrual methods of accounting?
7. What is the difference between the percentage-of-completion method and the completed-contract method of project accounting?
8. How is a balance sheet organized and what is it used for?
9. What is the difference between the current assets and the fixed assets?
10. What is the difference between the current ratio and the quick ratio?
11. How is working capital calculated, and why is it an important financial measure for a construction company?
12. How is the debt-to-equity ratio used in evaluating the financial condition of a company?
13. What are the three financial ratios that can be used to assess the profitability of a construction company? What are the differences among the ratios?
14. How would you develop a trend analysis for a construction firm?
15. What are the steps in performing a cash flow analysis for a construction company for the next 12 months?
16. How would you develop an annual operations budget for a construction company? How is it used to assess the financial performance of the company?
17. What are retained earnings and how do they affect the owners' equity of a company?
18. Why is it unwise for a construction company to finance the purchase of capital assets with its working capital?
19. What is a line of credit and why do many construction companies obtain one from a bank?
20. How would you establish an overhead budget for a construction company?

Exercises

1. Develop a chart of accounts for a construction company that primarily performs design-build projects. The company wishes to track income and cost by project and to keep separate records for design and construction services.
2. Conduct a detailed analysis of the operations budget shown in Figure 4.16 and develop a strategy for improving company profitability.
3. Prepare an assessment of the financial condition of Cascade Builders, using the data shown in Appendix B.
4. Prepare an assessment of the financial condition of Northwest Constructors using the data shown in Appendix C.

Sources of Additional Information

Fraser, Lyn M. and Aileen Osmiston. *Understanding Financial Statements*, 5th ed., Upper Saddle River, N.J.: Prentice-Hall, 1998.

Higgins, Robert C. *Analysis for Financial Management*, 7th ed., New York: McGraw-Hill/Irwin, 2004.

Industry Norms and Key Business Ratios, Bethlehem, Penn.: Dun & Bradstreet, published annually.

Jackson, Ira J. *Financial Management for Contractors*, 3rd ed., Raleigh, N.C.: FMI Corporation, 1999.

Milliner, Michael S. *Contractor's Business Handbook*, Kingston, Mass.: R.S. Means Company, Inc., 1988.

Palmer, William J., William E. Coombs, and Mark A. Smith. *Construction Accounting and Financial Management*, 5th ed., New York: McGraw-Hill, Inc., 1995.

Peterson, Steven J. *Construction Accounting and Financial Management*, Upper Saddle River, N.J.: Prentice-Hall, 2005.

Troy, Leo. *Almanac of Business and Industrial Financial Ratios*, Chicago, Ill.: CCH Inc., published annually.

CHAPTER 5
Strategic Planning and Management

INTRODUCTION

In the highly competitive construction industry, a construction company must plan for the future, if it intends to grow and remain a viable business enterprise. This involves forecasting the future business environment, evaluating alternative business strategies, and predicting potential future consequences of current business decisions. Based on this analysis, business strategies and action plans are selected to guide the allocation of resources and focus the efforts of the company. Strategic planning provides a company an opportunity to choose a future rather than be forced into one by reacting to the ever-changing marketplace.

Strategic planning involves the development of objectives and then linking them with the resources that need to be employed to attain the objectives. It attempts to develop answers to the following questions:

- What is our current situation?
- What do we want our future situation to be?
- What might inhibit us?
- What actions should we take to accomplish our objectives?

Strategic planning provides a mechanism for assessing company strengths and weaknesses, analyzing the external environment, forecasting the future, and selecting objectives and action plans that provide focus and guide company leaders in their pursuit of success. The primary purposes of strategic planning are to create a strategic management mind-set in every individual in the company, to reduce risk while improving profitability, and to improve decision making. Once a strategic plan has been crafted, it is used as the basis for managing the company.

Strategic management is the process of implementing the strategic plan and making adjustments, as necessary. It is an iterative process that involves

- monitoring and evaluating the external environment,
- assessing company internal capabilities,
- selecting collective company strategies,
- implementing the strategies,
- measuring company performance relative to the accomplishment of its strategic objectives, and
- reallocating resources, if necessary, to improve company performance.

CHAPTER 5 Strategic Planning and Management

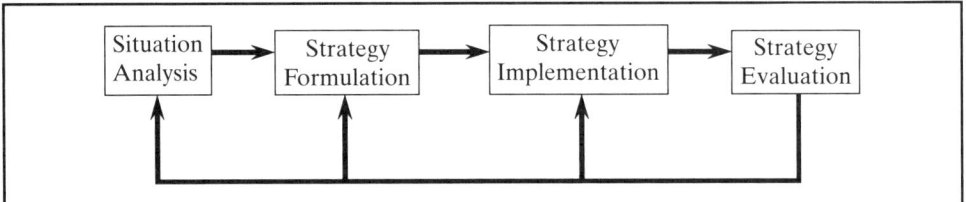

FIGURE 5.1 Strategic Management Process

The external environment may not evolve as forecasted during the planning process, and adjustments may need to be made to company strategies or action plans. This reassessment of planning decisions is illustrated in Figure 5.1.

There are three characteristics of strategic management. First, it is characterized by its emphasis on the interactions of the construction firm with its external environment. Throughout the planning and implementation processes, it is important that company employees continually forecast and assess how external factors may impact or are impacting company performance. The second characteristic is the focus on the interactions among the company's various functional activities, such as marketing, human resources management, financial management, and construction operations. The third characteristic is that strategic management concerns the choice of future direction based on the external and internal assessments. Performance is measured, and the company plan is adjusted, as necessary.

In this chapter, the various steps in strategic management are discussed. The first step in strategic management is situation analysis, which involves a strategic assessment of the external environment and the internal company capabilities. Next is strategy formulation, which is the development of specific company strategies. Once the strategies have been developed, they are implemented or put into action. The final step is strategy evaluation, which is the process of evaluating how the strategies have been implemented and their effectiveness in meeting company objectives.

Many construction firms have been reluctant to adopt strategic planning and management. Others have gone through the motions, but have not used strategic planning to its full potential. Company leaders and managers throughout the organization need a common understanding of the company, its markets, and its company strategies and programs for achieving success. The goal is to produce a plan that will be owned and understood by the people who have to execute it. This common understanding is the most important result of the strategic planning process.

Often, the planning process is more important to company success than are the products (plans) that come from the process. The analyses and brainstorming involved in developing the plan provide

- an assessment of the company's current posture,
- an assessment of the forecasted future environment opportunities and threats, and
- a set of strategies that will guide employees and managers in accomplishing their collective company vision.

A table of contents for a typical company strategic plan is shown in Figure 5.2, and an example of a plan is shown in Appendix A. Note that the strategic plan usually contains

CHAPTER 5   Strategic Planning and Management

Table of Contents

Strategic Assessment
    External Analysis
    Internal Analysis
Strategic Direction
    Mission Statement
    Corporate Vision
    Strategic Objectives
    Strategies to Accomplish Objectives
    Action Plans to Execute Strategies
Financial Plan
Business Development Plan
Resource Plan
    Capital Assets
    Human Resources
    Information Management
Contingency Plan

FIGURE 5.2 Typical Construction Company Strategic Plan

supporting plans, such as financial, business development or marketing, resource, and contingency plans. Strategic planning requires the commitment of company decision makers and buy-in by all those involved in the plan's implementation. Without it, the results may be marginal, and the entire effort compromised.

Managing strategically means making decisions and implementing strategies that allow a construction company to develop and maintain a competitive advantage. What do we mean by competitive advantage? It is some service characteristic that sets the company apart from its competition. It may be the ability to provide services with unique characteristics—such as superior performance or quality, cost or timeliness—or the ability to develop superior relationships with customers. Both forms of competitive advantage require an understanding of customers' needs and of what creates value for them. To have a competitive advantage, the company must possess unique organizational assets and capabilities. The capabilities come from the company's core competencies and its focus on customer satisfaction. Customer expectations change over time, as do the capabilities of the construction firm's competitors. Understanding these market shifts and selecting appropriate strategies are the necessary ingredients for a construction company's long-term success.

PLANNING PROCESS

Figure 5.3 provides an overview of the planning process from the creation of a mission statement through the establishment of performance measurement criteria. Each step in the process is described in this section. The objectives of the process are

- to establish a framework for decision making,
- to assign responsibilities and allocate resources, and
- to establish quantifiable goals that can be used to measure company progress.

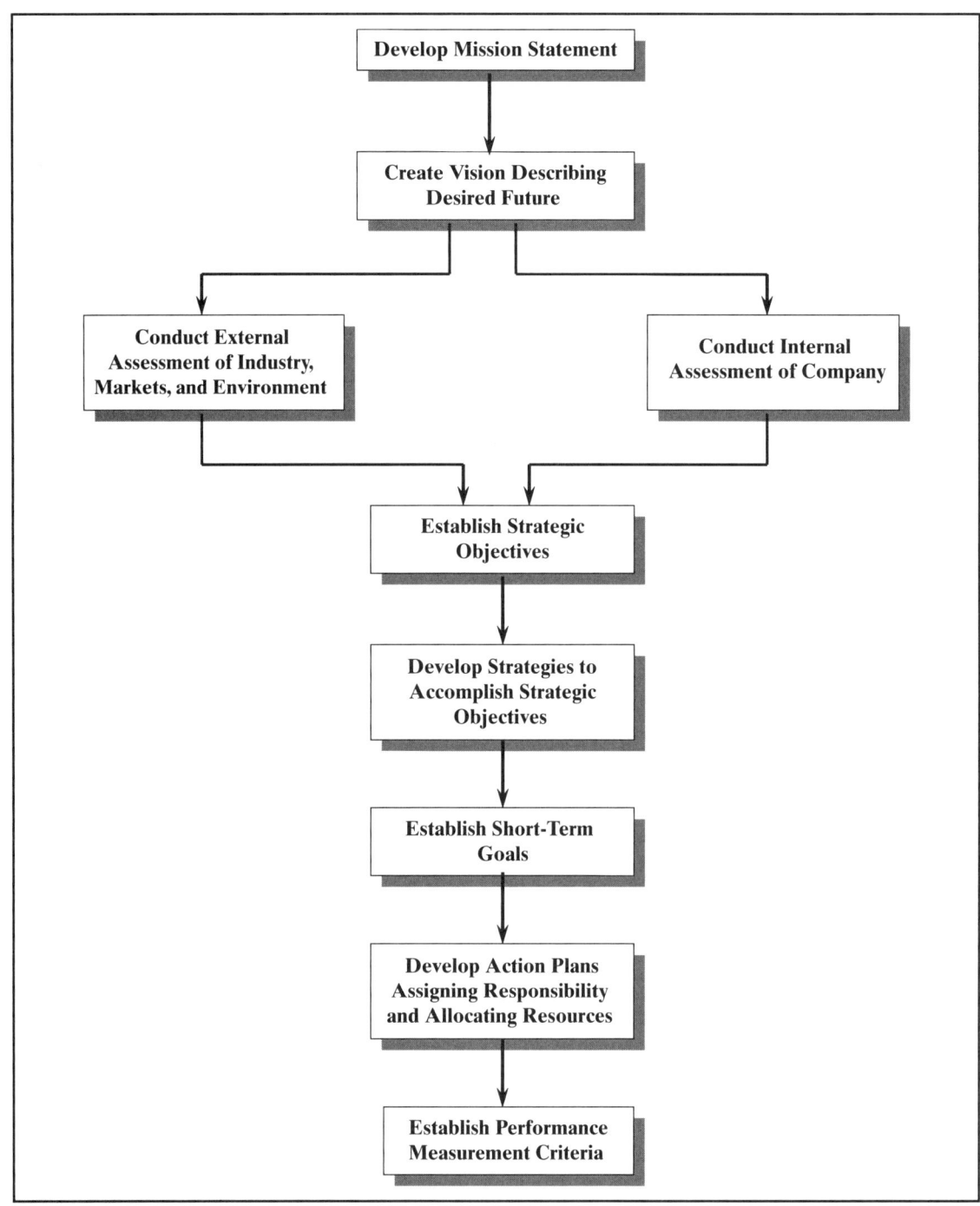

FIGURE 5.3 Strategic Planning Process

CHAPTER 5   Strategic Planning and Management

The result is an integrated plan that can be used to guide decision makers in managing the company.

Mission Statement

The initial step in developing a strategic plan for a firm is to develop a meaningful mission statement that has significance to both the employees and the customers of the firm. The **mission statement** should articulate in simple terms why the company exists and what makes it different from other firms in the industry. This statement should be carefully crafted

- to be of enduring value,
- to provide focus to the organization, and
- to establish boundaries for the development of strategic objectives and goals.

A cross-section of company employees should be used to develop a meaningful mission statement. This ensures that different perspectives are considered in crafting the statement and helps ensure its acceptance at all levels within the company. Facilitators are often used to help in the development of mission statements.

Some examples of company mission statements are shown below.

> We build quality homes and neighborhoods that exceed the expectations we have established with our customers.

> Great Northern Constructors is an enthusiastic, innovative, client-centered construction services organization, committed to exceeding our clients' expectations in an atmosphere of trust and respect. Our purpose is to provide unsurpassed service, risk management, quality and value, while achieving profitability through competitive pricing and creativity. Our people are the essence and vitality of our company. Together in partnership, we create a safe, highly productive, and enjoyable environment, which challenges and enriches each individual. Collectively, we provide the opportunity for personal and professional growth in support of individual plans and goals.

> Western Construction constructs high-quality, management-intensive construction projects in the Western States and provides professional, client-oriented services. Our clients perceive us as a customer-oriented organization that delivers value better than the rest of the industry. We are a profitable enterprise that undertakes only projects that have the prospect of an appropriate financial return. We have a strong, cohesive team of talented individuals with high esteem and self-worth. We provide a work environment that demands respect for every individual, promotes professionalism, is upbeat, and is a fun place to work. We are a responsible corporate citizen, making contributions in the community and respecting the environment.

> International Constructors is an innovative leader of the industrial construction industry committed to the total service of each client. This is accomplished by producing quality projects on time at competitive prices in partnership with our clients, design professionals, trade leaders, craftsmen, and suppliers. We take pride in
>
> - Our individual commitment to the total satisfaction of the client.
> - Our skill in assembling specialized project teams.
> - Our integrity, prosperity, and growth.

- Our ability to be at the forefront of an ever-changing industry.
- Our ability to understand the specialized requirements of our clients.
- Our commitment to promoting feelings of high esteem in our people.
- Our emphasis on providing a safe working environment on all projects.

These four mission statements each carry a message. They incorporate the company's self-image and philosophy. They represent firms of varying size and market focus within the construction industry, and yet, each has a common thread. What do these mission statements say about the core values of each company?

Company Vision

Once a mission statement has been developed, a **vision** statement is written that broadly describes where the company intends to be in the next ten or so years. This vision should be

- succinct,
- inspiring to employees, and
- meaningful to current and potential customers.

This vision will serve as the foundation for the development of strategic objectives. A successful vision provides employees the opportunity to view their positions as members of a team that has a unified direction and purpose, rather than viewing their roles as being responsible only for their own individual tasks. The vision guides decision making within the company and aligns the activities of employees so that they work together as a cohesive team.

As with the mission statement, a cross-section of company employees should be used to develop a meaningful vision for the firm. To ensure its acceptance at all levels within the company, it is suggested that the final draft be sent to all employees for comment. Based on employee feedback, the final version is crafted. While the mission statement is written using present tense because it describes what the company does, the vision is written to describe what the company aspires to become in the future.

Some examples of company vision statements are shown below.

> To become a national home-building company with the financial and organizational strengths to provide quality homes in distinctive neighborhoods throughout the nation.

> Western Construction will be the premier building contractor in the Western States known for its ability to
> - Create continuing partnerships with high-quality owners to produce best-value construction projects.
> - Recruit, train, and retain the best people in the industry at all levels within the organization.
> - Self-perform key elements of the work with highly productive, skilled craftsmen.

> To become the constructor of choice in the Southeastern States by creating job satisfaction and adequate rewards for our employees, building quality projects for our clients, and becoming recognized within our industry and the communities we serve as an outstanding organization.

These three company visions have much in common. They each state the role the firm aspires to play within the industry and provide focus and direction to employees. The vision displays the company's attitude and intentions to its existing and potential customers, its current and future employees, its competitors, and to the general public. A properly crafted vision captures the essence of a company.

Strategic Assessment

Once the mission statement and company vision have been developed, an assessment of the existing company posture is made to serve as a baseline for the development of the strategic plan. The assessment needs to identify company strengths that can be capitalized upon in developing action plans, as well as company weaknesses that must be addressed. By fully understanding company strengths and weaknesses, company leaders will be able to make realistic decisions regarding business initiatives and allocate resources to correct significant weaknesses. This internal assessment will be discussed in more detail in the section "Situation Analysis."

An **external assessment** is made to identify opportunities for future successes and potential threats to the company. This external assessment should evaluate the company's current and potential customers to understand their selection criteria and motivation to purchase services. Are there unmet needs among the customer base? The external analysis needs to carefully analyze existing and potential competition. What will be the bargaining power of customers or material suppliers in the future? What will be the impact of technology, government, demographics, culture, and economics in the future?

This external assessment should identify all external factors that will significantly impact the company's future. In most cases, the future performance of these factors needs to be forecast based on probable external events. The output of the external assessment is a forecast of the future environment and its impact on the company and the consequences of continuing to do business as it is being done today. This forecast may be based either on statistical projections of historical data or on brainstorming by knowledgeable individuals. External assessments will also be discussed in more detail in the section "Situation Analysis."

Strategic Objectives

Strategic objectives are developed to guide the company in the accomplishment of its vision. These objectives should be broad and timeless statements that require action. Successful accomplishment of the objectives should take the company from the baseline determined in the strategic assessment to the posture described by the company vision. The strategic objectives should address both financial and nonfinancial issues. They should be realistic, yet challenging, and stretch the organization. The objectives must be measurable and should incorporate a time dimension, so progress can be monitored. They must be structured to avoid conflict among elements within the company and any unintended consequences.

These objectives need to be specific and action-oriented, so there is a clear understanding regarding what is to be accomplished. Does the company want to increase its profitability or market share? Maybe the company wants to diversify into other markets. Measurable objectives are important because in order to get anywhere, a company must decide where it wants to go and have a way of knowing when it has arrived. Abstractions and generalizations are not enough. Development of strategic objectives will be discussed in more detail in the section "Strategy Formulation."

Company Strategies and Short-Term Goals

Company strategies are selected to describe the methods to be employed to accomplish each objective. These strategies define both what is to be done and how it is to be done. The selected strategies should be able to utilize the company's core capabilities to exploit the forecast external opportunities while minimizing the risks associated with anticipated threats. One way to proceed is to list each of the major alternative strategies for objective accomplishment. Each alternative strategy should be evaluated in terms of specific risk factors under alternative future forecasts. This allows a sensitivity analysis of strategy selection with respect to possible alternative futures.

Strategies, including human resource and capital investment, are developed for accomplishment of the strategic objectives. Strategies must take into account the human and financial resources of the company realistically. Balancing risk with potential reward in light of current realities is the challenge in the development of corporate strategies. Planning assumptions made in developing external forecasts need to be reviewed during the development of corporate strategies to minimize risk to the company in the event that external factors do not materialize as forecast. These assumptions also need to be reviewed during the performance evaluation session to ensure that the assumptions are still valid. Development of company strategies will also be discussed in more detail under "Strategy Formulation."

Short-term goals are selected as intermediate targets toward accomplishment of long-term strategic objectives. While the planning horizon for a strategic plan is typically five to ten years, the planning horizon for short-term goals is one to two years. Strategic plans should be reviewed annually, and new short-term goals selected for the succeeding one to two years. These short-term goals are used for developing action plans that assign responsibility and allocate resources.

Action Plans

An **action plan** is an assignment of specific actions to an individual, an allocation of resources, and an establishment of a specific date by which the actions are to be completed. Detailed action plans are developed assigning responsibility for goal accomplishment. These action plans should be developed for the major business functions within the company, such as marketing, financial management, project execution, human resource development, and information management. These action plans need to specify which individuals will accomplish what tasks by what date with which resources (people and money). The assignment of responsibility is critical to the implementation of the strategic plan. The action plans must be clearly communicated to those responsible, and the required resources must be allocated. Individuals who are assigned responsibility for action plan accomplishment must understand the significance of plan accomplishment to the achievement of overall company strategic objectives. Development of these action plans will be discussed in more detail under "Strategy Implementation."

Performance Measurement

Once the action plans have been developed, a feedback structure needs to be implemented to assess progress toward accomplishment of company goals. Since it costs money to collect data, it is important not only that sufficient data be collected to make a reasoned analysis of company progress, but also that unnecessary data not be collected. Data collection must include some form of feedback from the company's customers to ensure satisfaction with the services received. The most common feedback system is a monthly or a quarterly review and analysis of company performance.

Specific data are collected relating to business volume, profitability, cost of doing business, customer satisfaction, marketing initiatives, and employee development. The data are reviewed by company executives to assess progress toward meeting company goals and objectives. When the reviewed data indicate that progress is not being made, the executives need to determine the reasons. It may be that external conditions did not occur as forecast in the external assessment, it may be that insufficient resources were allocated to the responsible parties, it may be that the goal was too optimistic, or it may be that a different set of strategies are needed. Based on this analysis, parts of the strategic plan may need revision, as discussed earlier. This is where a feedback mechanism is required in strategic management.

Now let's examine the four major steps in strategic management.

SITUATION ANALYSIS

Before selecting an appropriate set of company strategies, the company needs to analyze the current situation. This strategic assessment is sometimes referred to as conducting a **SWOT** analysis. The **S** and **W** refer to assessing the *strengths* and *weaknesses* of the company, and the **O** and **T** refer to forecasting future *opportunities* and the potential *threats* the company may experience. The reason for developing the mission statement and the vision before conducting the situation analysis is to provide a context for assessing the external environment and the company's internal capabilities. Strategic planning is concerned only with those issues that relate to the company's ability to accomplish its mission and vision.

External Analysis

An external analysis is the process of scanning and evaluating the external environment to determine positive and negative trends that could impact the company's performance. This is the process of forecasting future opportunities and threats. Opportunities are positive external trends that will help the company improve its performance, while threats are negative trends that could hinder its performance. During the development of a strategic plan, it is important to know what is happening in the external environment and to forecast future trends so that strategies can be selected to take advantage of the opportunities and to avoid or mitigate the impact of the threats.

The external environment is the source of information, the source of customers, and the source of work for a construction company. Information needed includes the anticipated expectations of potential customers, the anticipated service characteristics of competitors, and the availability and cost of resources (materials, equipment, labor, and capital). To sustain a competitive advantage, company leaders need to recognize and anticipate environmental changes and understand their potential impacts on the construction firm.

The environmental analysis should be performed at two levels, the general *business* environment and the specific *industry* environment. Both have significant impacts on the company's ability to generate business and profits.

General Business Environment

The general business environment consists of those environmental sectors that have an indirect effect on the construction company's strategic decisions and actions, but provide a context within which the specific industry environment must function, as shown in Figure 5.4. The general business environment sectors include the economic, the demographic, the political, the technological, and the sociocultural environments, as illustrated in Figure 5.5. Changes in these sectors may create opportunities that the construction firm can exploit or may pose threats to its current business volume. Each will be discussed in the following paragraphs.

The **economic environment** includes the economic factors that affect the customers' ability to purchase construction services and the construction firm's cost of doing business. These include interest rates, exchange rates (if any business is denominated in foreign currency), inflation rates, economic growth or recession, and government budget deficits or surpluses. What one is looking for in analyzing these statistics are current values and forecasted trends. One needs to predict what impacts, if any, these trends may have on one's company's business. For example, a declining interest rate may make construction more affordable for prospective customers. One also needs

FIGURE 5.4 Construction Company's External Environment

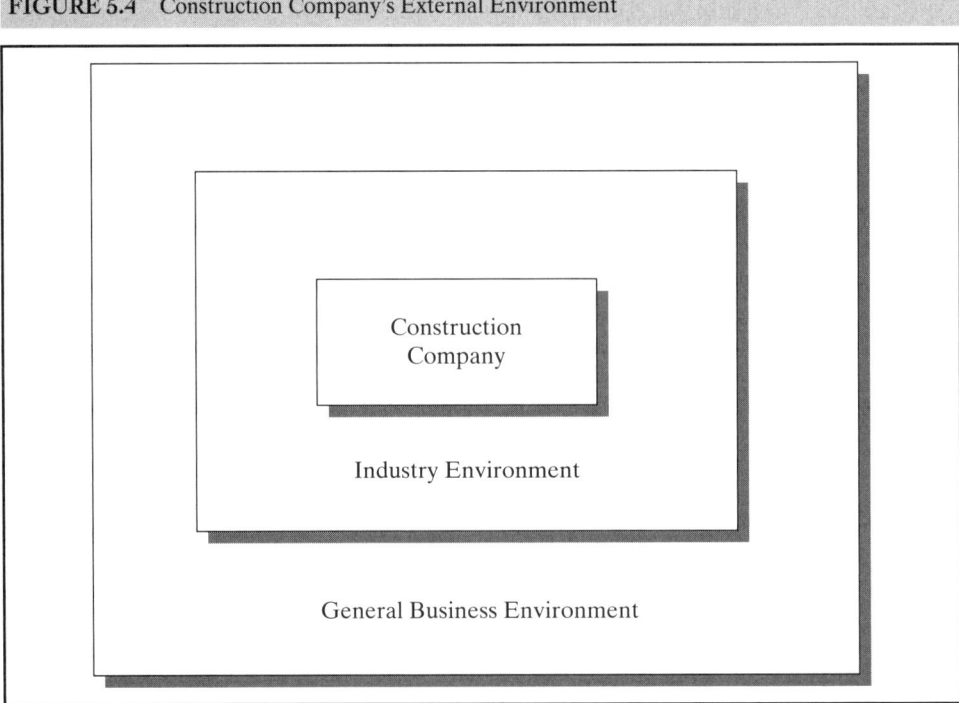

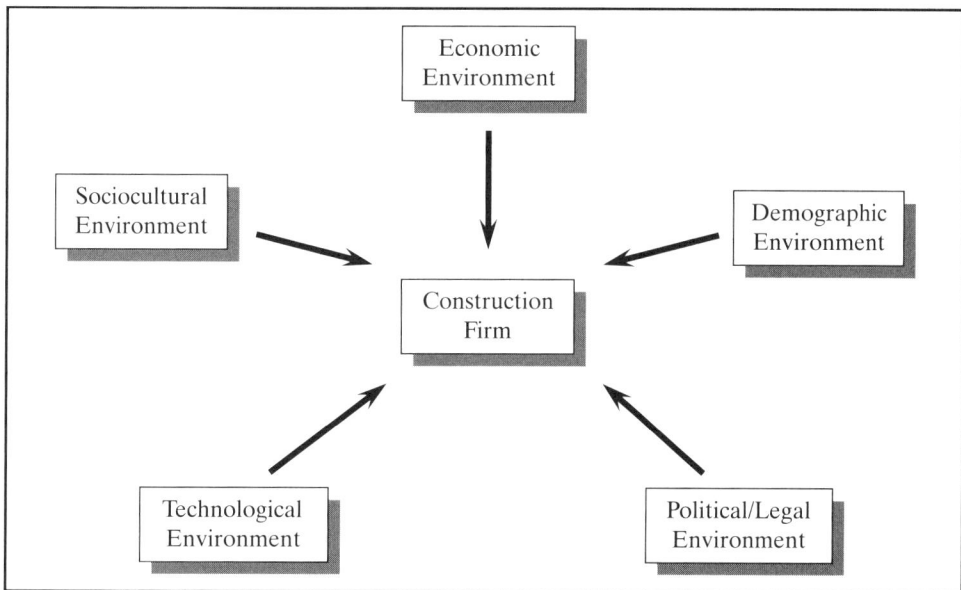

FIGURE 5.5 General Business Environment

to forecast the potential impact of forecasted trends on the demands for customers' products and services, as this will influence their decisions regarding whether or not to develop additional construction projects. For example, a recession generally results in reduced customer demand, and retail merchants may defer construction of new stores and distribution centers.

The **demographic environment** includes the age distribution of the population, its ethnic composition, the regional changes in population growth, the employment rates, the education levels, and the availability of prospective new employees. Most of the current and historical statistical data can be obtained from the U.S. Census Bureau. A census of the construction industry is taken every five years. As one looks at population statistics, one should examine historic trends and make future forecasts for those that potentially may have the greatest impact on one's company. This type of information is useful in understanding one's current customers and for targeting potential customers. For example, if a company specializes in constructing assisted-care living facilities, an aging population would indicate an increased demand for their services. An aging workforce would indicate a need to expand the pool of people from which to recruit and a need to market careers in the industry to them.

The **political/legal environment** includes government expenditures, laws, regulations, judicial decisions, and political forces. These factors may significantly affect the company's markets (if pursuing public projects) and cost of doing business (government regulation). This includes an analysis of actions at all levels of government—federal, state, and local. In analyzing the impact of government regulation, one needs to forecast anticipated changes during the period being considered and to examine the potential impact of these forecasted changes on the company's way of doing business. Taxation policies may also change over time and impact the cost of doing business for the construction firm.

The **technological environment** includes technological changes and trends that may impact construction operations or construction management, as the use of cellular telephones have had since the beginning of the 21st century. Construction projects are becoming more sophisticated, and electronic documents are being used to transmit both contract drawings and specifications. Web-based collaboration tools are increasingly being used to manage construction projects. The challenge is to get designers, subcontractors, and suppliers to participate in the use of these technologies. The major concern is to forecast the technological development in construction components and techniques as well as in construction management processes in the next five to ten years. This may require brainstorming sessions to predict future technological advances. Future developments may change the way in which we conduct construction business, just as e-mail has largely replaced written communications.

The **sociocultural environment** includes cultural values, attitudes, behaviors, and opinions. One wants to know how society values, attitudes, and patterns of behavior are changing. This information tends to be quite subjective, but is critical when planning for the future. People's attitudes influence their behavior at work and their interest in working for a specific company. The challenge is how to attract quality people and how to keep them motivated to remain as productive members of the company team. This also relates to the interest of younger employees in the use of sophisticated technology in their work. Another major issue for construction company leaders is the general public's opinion of careers in the construction industry and how it is likely to change in the future.

Specific Industry Environment

The specific industry environment consists of those environmental sectors that have a direct and immediate impact on a construction firm's decisions and actions by presenting opportunities or posing threats. These sectors include customers, suppliers, existing competition, new market entrants, and subcontractors, as illustrated in Figure 5.6. Each will be discussed in the following paragraphs.

Customer analysis involves identifying the company's major customers and potential future customers, the characteristics of the firm's services that are likely to be most important to existing and prospective customers, and the customer needs that are not being satisfied currently. What is the forecast future demand for the type of services that the construction firm offers, and who will have leverage in the market, the customers or the construction firm? A growing market may provide leverage to the construction firms as demand for construction services may exceed the capabilities of local construction companies. A shrinking market, however, usually provides leverage to the customers as construction capacity may exceed demand. Who are the firm's largest customers? What changes in customer-purchasing motivation are anticipated? Are customers moving to value-based procurement decisions rather than solely cost-based? Are they going to seek integrated design and construction services? What service characteristics will be most important to existing and prospective customers?

Subcontractor analysis involves assessing the future availability of financially sound quality subcontractors who employ skilled craft workers. Who will have leverage in the market, the subcontractors or the general contractors? What is the relationship between desired subcontractors and the construction company? If subcontractors are not treated well, they may be unwilling to participate in a project if there is more demand for their services than the available supply in the market. If demand exceeds

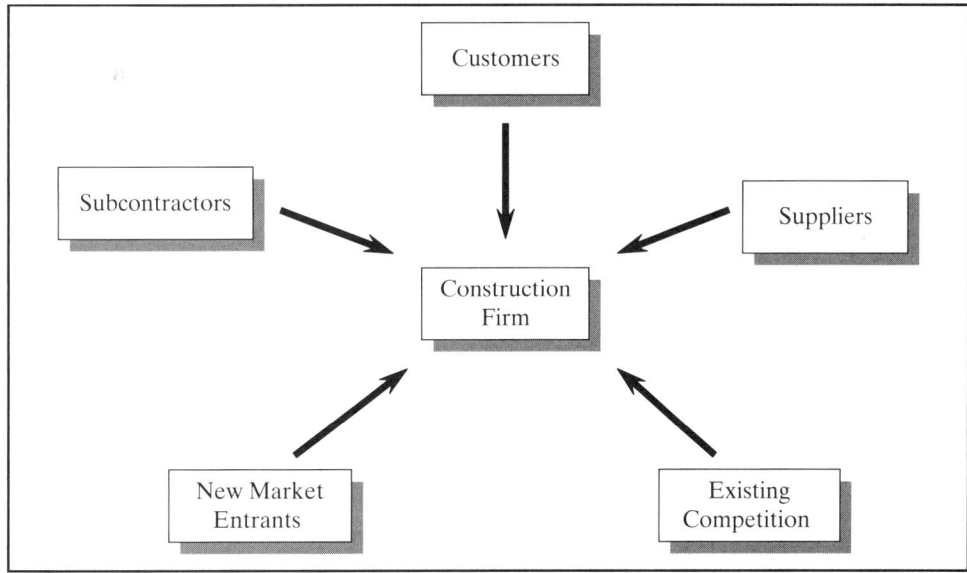

FIGURE 5.6 Specific Industry Environment

supply, the subcontractors have more control over the price of their services, which will affect a general contractor's cost of doing business.

Supplier analysis involves evaluating the future availability of reliable suppliers who provide quality products at competitive prices. Who will have the leverage, the buyer or the sellers of construction materials? Supplier analysis is similar to that for the subcontractors. What is the availability of materials compared to the demand? If structural steel is in short supply, the suppliers will have more leverage regarding price and may be more selective in supporting construction companies that have a reputation for paying for materials when invoiced. Understanding the suppliers will help in crafting company strategies for developing good supplier relationships.

The *competitor analysis* involves assessing the threat of new entrants into the market and the future capacity of current competitors. The existing competitors are those other companies that offer similar services to the same set of customers that one's firm serves. The purpose of this analysis is to determine who they are and what services they offer. Are the competitors growing by increasing their market share? If so, what is the source of their competitive advantage? What new services are being offered by competitors, and how are customers responding? A growing market may attract new entrants. What is the cost of entering the market, and what is the most likely means of entry? New competitors may establish a new presence, purchase a company, or form a joint venture with an existing firm.

Use of Analysis Results

The objectives of the external analysis are to forecast the future business environment for the firm and to identify potential opportunities and threats. This provides a context or background for the development of the strategic plan. The following is an example of an external analysis for a general contractor that specializes in commercial construction.

Analysis of General Business Environment

Economic environment. Financing and commercial lending interest rates will increase to control inflationary pressures due to the growing economy. The rental/lease rates for commercial office and retail space are at high levels and are expected to increase by 5 percent over the next five years.

Demographic environment. The regional business forecast for the next five years is positive. Eighty percent of the employers in our market area report that they expect revenue growth and additional demand for workers over the next five years. Overall job growth is expected to average 2 percent, and population growth is expected to average about 3 percent over the next five years. Demand for commercial construction is expected to remain high because of the anticipated population growth. Construction labor will be in short supply and will impact the cost of construction.

Political/legal environment. Environmental concerns, attempts to limit the spiraling cost of construction, and concerns regarding maintaining and improving public infrastructure will result in increased governmental regulation of the construction industry and will complicate the permitting process. This trend could result in increased construction costs and place some downward pressure on the demand for commercial construction.

Technological environment. Building systems will become more sophisticated, particularly in the electrical and mechanical areas. More environmental-friendly building components will be developed, and there will be an increased use of composite materials. Web-based collaboration tools will be required by some customers, necessitating the training of company employees and major subcontractors.

Sociocultural environment. There will be an increased emphasis on reducing the environmental impact of construction. High school graduates and their parents will continue to undervalue careers in the construction industry.

Analysis of Specific Industry Environment

Customer analysis. The company has a broad customer base, and the growing economy should provide opportunities for increased work. Customers are becoming more demanding and expect the general contractor to assume increased responsibilities for managing the permitting and design processes. Increased customer interest in the design-build delivery method is expected over the next five years.

Subcontractor analysis. Increased demand for quality subcontractors will increase their selectivity regarding the general contractors with whom they wish to work and provide subcontractors greater leverage in pricing their services.

Supplier analysis. The cost of construction materials and supplies is expected to increase by 4 percent per year over the next five years. Local high demand for supplies may result in supply shortages during the summer construction periods. These shortages will increase the bargaining power of suppliers.

Competitor analysis. The commercial construction market is highly competitive and increased market demand may result in new entrants or diversification of existing competitors. Current competitors may expand their service offerings to increase their market share and meet increased customer expectations.

Internal Analysis

To be able to select appropriate company strategies, it is important to determine what the company can and cannot do well and what assets it has and does not have. This is called an internal analysis or **internal assessment** of company strengths and weaknesses. It serves to define the existing conditions that are needed to establish a baseline for the development of a strategic plan. Strengths are the resources that the company possesses and the capabilities that it has developed, which can be exploited to develop a sustained competitive advantage. Weaknesses are resources or capabilities that are lacking and prevent the company from developing a sustained competitive advantage.

An internal analysis is the process of identifying and evaluating the company's resources, capabilities, and core competencies. Company resources include financial, physical, human, intangible, and organizational resources.

- Financial resources are receivables, cash, demand deposits, certificates of deposit, and other financial instruments owned by the company.
- Physical resources tend to be facilities and equipment.
- Human resources include the experiences, knowledge, skills, accumulated wisdom, and characteristics of the firm's employees.
- Intangible resources are the company's reputation for quality work and customer service.
- Organizational resources are the corporate culture, company business processes and policies, corporate relationships, and organizational structure.

Company capabilities are the organizational processes that determine how effectively and efficiently customer needs are met. **Core competencies** are the company's major value-creating skills and capabilities. They establish what the firm does best and where its strength resides. A company's organizational capabilities and core competencies are the foundations upon which competitive advantage is developed and sustained.

There are three basic approaches that may be used in conducting an internal analysis—a value chain analysis, an internal audit, and an internal functional analysis. Each approach is described in the following paragraphs. It is often desirable to use all three approaches.

Value Chain Analysis

A **value chain analysis** is a systematic approach to examining all of the company's functional activities and assessing how well they create customer value. By assessing a firm's strengths and weaknesses with respect to each of these activities, managers can develop an in-depth understanding of the company's capabilities. Creating customer value is why the company exists, and strategic planners need to understand which values are important to customers. Identification of customer expectation is also a key element of the external analysis discussed previously.

The basic premise behind value chain analysis is that customers expect to receive value from the services that the construction firm provides. The analysis is performed by assessing how value is produced within the company's business processes. Each process not only contributes to the creation of customer value, but also has a cost. Does the increase in customer value generated by the process exceed the cost of performing the process? Value-creating activities can be divided into two major categories—primary activities and support activities.

The primary business activities of a commercial construction company are creating relationships with prospective customers and obtaining the work (business development), procuring materials and subcontractors, constructing the projects, and processing warranty issues or handling other post-construction services. Supporting activities include the basic company organizational structure, its employees and equipment, as well as human resources management programs and policies to attract, develop, and maintain a motivated, skilled team of employees. This is illustrated in Figure 5.7.

A similar value chain for a residential builder is shown in Figure 5.8. The primary business activities are acquiring the land for developing a neighborhood of new homes, developing the site by installing the street network and utility systems, designing the various models of homes to be offered to prospective customers, procuring materials and specialty contractors to construct the homes, managing the construction process, marketing the homes to prospective buyers, and handling any warranty or other post-construction issues. The primary support activities are the company infrastructure and its human resources management programs.

Each of these activities can be further divided for conducting a thorough internal analysis. Each item is then rated as poor, below average, average, above average, or excellent to create an overall assessment. An example of a breakdown for the primary activities shown in Figure 5.7 is as follows:

Business Development

- Effectiveness of market research in identifying customer needs
- Motivation and effectiveness of business development staff
- Loyalty of current customers
- Ability to attract new customers
- Dominance of company in desired market sectors

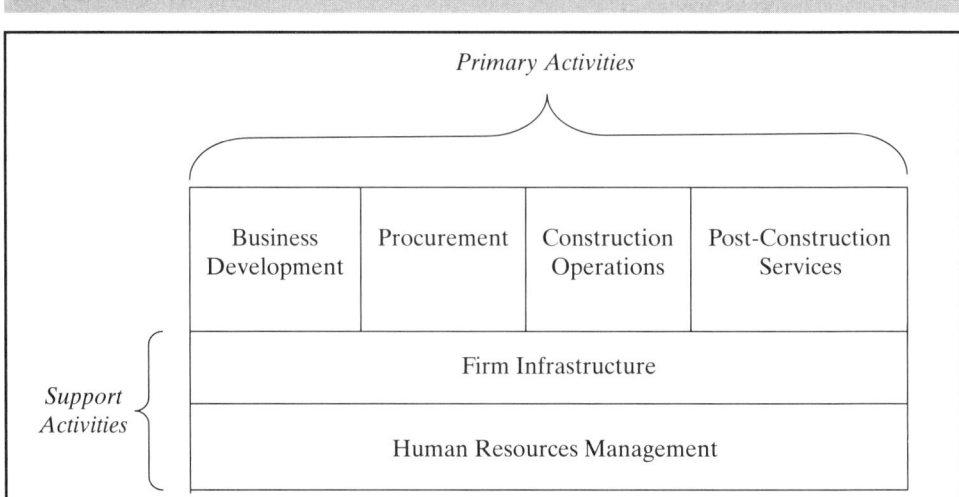

FIGURE 5.7 Value Chain for Commercial Contractor

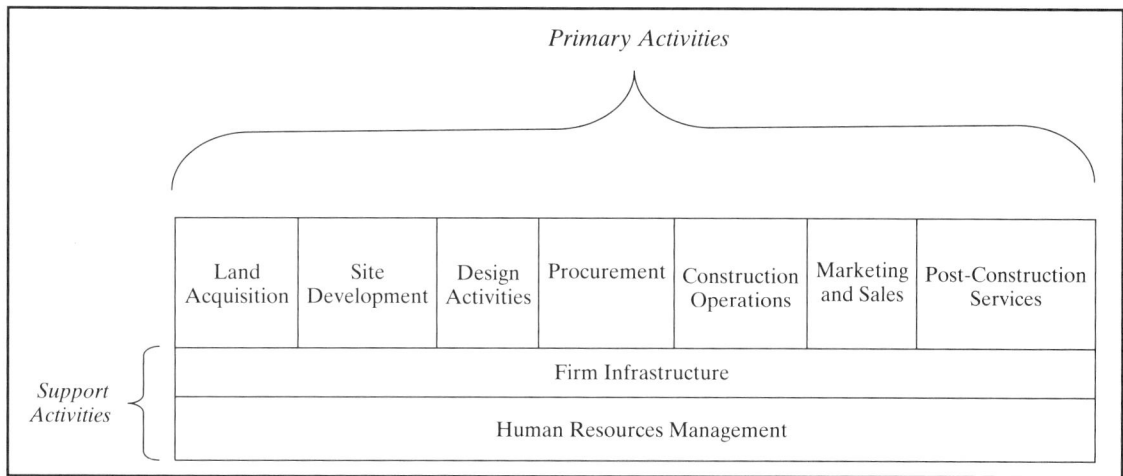

FIGURE 5.8 Value Chain for Residential Builder

Procurement

- Relationships with subcontractors
- Efficiency and effectiveness of subcontracting procedures
- Relationships with suppliers
- Efficiency and effectiveness of material purchasing procedures
- Efficiency and effectiveness of submittal management procedures

Construction Operations

- Effectiveness of cost-estimating and cost control procedures
- Effectiveness of project scheduling and schedule control procedures
- Effectiveness of project management systems
- Effectiveness of quality management systems
- Effectiveness of safety management systems

Post-Construction Services

- Effectiveness of warranty management system
- Effectiveness of customer relationship development program

The support activities also can be further divided for detailed assessment as shown below:

Firm Infrastructure

- Adequacy and location of facilities and equipment
- Efficiency and effectiveness of finance and accounting system
- Efficiency and effectiveness of information management system
- Efficiency and effectiveness of business management systems
- Effectiveness of overhead cost control system

Human Resources Management

- Effectiveness of procedures for recruiting and developing employees
- Appropriateness of employee reward system
- Working environment
- Relations with unions
- Levels of employee motivation and job satisfaction

A value chain analysis provides an assessment of the company's capabilities and business processes from the customers' perspective. The analysis helps identify company strengths and areas that need improvement.

Internal Audit

A second approach to conducting an internal analysis is to conduct an **internal audit**. This involves assessing the company's major functional areas to determine if they have adequate resources to perform their work activities and how well they perform their assigned tasks. The primary functional areas of a general construction company that should be audited are marketing, financial management, contract management, cost estimating, construction management, procurement, equipment management, human resources management, and information management. Some key audit questions are listed below for each of these functional areas.

Marketing

- What is the effectiveness of the company's marketing activities?
- What is the company's market share?
- What marketing tools are being used?
- How strong are the company's relationships with major customers?

Financial Management

- What is the company's current financial condition?
- Does the company maintain adequate working capital?
- What are the efficiency and the effectiveness of the company's accounting procedures?
- Does the company have an effective overhead budgeting and cost control system?

Contract Management

- Does the company have effective procedures to maintain all contract documents?
- Does the company have effective procedures to manager contract change orders?

Cost Estimating

- Do project cost estimates compare favorably with final project costs?
- Does the company maintain a cost-estimating database that is updated with historical project cost data?
- Does the company have effective procedures for estimating the cost impacts of change orders?

Construction Management

- Does the company have adequate quality control procedures to meet customer expectations?
- Does the company deliver projects on time?

- Does the company have effective safety programs and a good safety record?
- How does the project team manage project documentation?

Procurement

- What are the company's relationships with subcontractors?
- How effective are subcontract procurement procedures?
- What are the company's relationships with suppliers?
- How effective are material procurement procedures?

Equipment Management

- Does the company equipment fleet generate a profit for the company?
- How are acquisition and disposal decisions made?
- Are realistic rates used to charge projects for the use of company-owned equipment?
- Does the company have an effective equipment maintenance program?

Human Resources Management

- What is the working environment within the company?
- What is the employee turnover rate in the company?
- What is the absentee rate among employees within the company?
- What employee training and development programs are available within the company?
- What type of reward system is used by the company?
- How effective are company recruiting efforts?
- What is the relationship between the company and the craft unions?

Information Management

- How is information managed within the company?
- Does the company use common databases to capture and manage company data?
- What paper or electronic files are retained and who manages them?
- Is information delivered efficiently and effectively?

Internal Functional Analysis

The third approach to determining a company's strengths and weaknesses is to conduct an **internal functional analysis**. This involves identifying and evaluating the firm's resources, capabilities, and core competencies. Company resources include

- physical assets,
- financial assets,
- human resources,
- reputation, and
- organizational assets.

The physical assets of a construction firm tend to be facilities and equipment. These assets are evaluated based on their location and capability to meet competitive challenges. Financial assets are the financial resources available to finance the operation of the firm.

Human resources are the number and skills of the firm's employees. Reputation has two major components; the company's reputation as a business enterprise within the community and the company's reputation for delivery of quality construction

- Dedicated, hard-working, smart, and cooperative people
- Strong commitment to self-performed work
- Safe, efficient, and well-equipped crews
- Strong value-driven business
- Regional recognition for outstanding work
- Affiliation with quality owners with whom we enjoy doing business
- Strong team of partners (subcontractors, suppliers, bonding, banking, and legal)
- Geographic flexibility

FIGURE 5.9 Sample Set of Core Competencies

services. Organizational assets are the corporate structure, company culture, and business processes.

Company capabilities include

- Management capabilities, such as leadership, decision making, and efficient use of company assets
- Use of financial assets in terms of liquidity, leverage, and profitability
- Technical skills of employees
- Teamwork skills of employees
- Capital and overhead budgeting procedures
- Information management system
- Human resources management system
- Business development procedures

Each company resource and capability is evaluated and rated as poor, below average, average, above average, or excellent. This will lead to identification of the company's core competencies, or major value-creating skills and capabilities. Do these core competencies provide the company a competitive advantage in the market? If so, they should be emphasized during the development of strategic objectives. If not, they need to be supplemented with additional capabilities. An example of a set of core competencies for a construction firm is shown in Figure 5.9.

Use of Analysis Results

The objective of the internal analysis is to identify the company's strengths and weaknesses. This provides a baseline of the firm's current posture that will be used in selecting the strategic objectives and the company long-range strategies for their accomplishment. The goal is to move the company from its current position as described by the internal analysis to the position described in the company vision.

STRATEGY FORMULATION

To be successful, a construction firm must hold some competitive advantage relative to its competition. This advantage must relate to one or more attributes of the scope of services provided. One of the objectives of strategic planning is to identify the criteria prospective customers use in selecting a construction firm and to develop the core competencies that allow the services provided to differentiate the company's capabilities from those of its competition. This differentiation may be in the form of low cost, high quality, technological innovation, post-construction services, or quick response.

The situation analysis previously discussed provides a baseline for the development of the strategic plan. The analysis provides an assessment of the company's strengths and weaknesses and a forecast of the firm's future business environment. We now return to the questions posed in the introduction to this chapter. What actions should be taken toward accomplishment of the company vision? First, long-term strategic objectives are established, such that if accomplished, the company would accomplish its vision. These objectives are measurable statements of desired outcomes. They should be specific, feasible, and tangible. The reason for using quantifiable objectives is to be able to measure progress toward their accomplishment. Subjective objectives are almost impossible to measure and leave too much opportunity for different interpretation.

Company long-term objectives typically address the following issues:

- Market standing
- Innovation
- Productivity
- Physical and financial resources
- Profitability
- Managerial performance and development
- Worker performance and development
- Public responsibility

The following are some examples of strategic objectives that a construction firm might select:

- Decrease long-term debt by $250,000 per year over next ten years.
- Increase company profitability by 3 percent per year over next ten years.
- Increase revenues by 5 percent per year over next five years.
- Expand market into new geographical areas.

Once strategic objectives have been established, strategies are selected for accomplishing the objectives. These strategies describe both what is to be accomplished and how it is to be done. They must be compatible with the company structure and the firm's core competencies. Care must be taken when selecting strategies to identify any assumptions being made regarding the future business environment. This is to highlight the need to reconsider strategies in the event that the future business environment is significantly different from the one forecast. Three sets of company strategies are developed—business, competitive, and functional strategies. The relationship among these three sets of strategies is shown in Figure 5.10. The strategies should focus on developing or enhancing the company competencies and the capabilities for improved customer value that will result in a sustained competitive advantage.

The first set of strategies to be developed is the *business strategies*. They address the desired mix of business for the company and the desired direction in which the company is going. Business strategies tend to address such issues as market share, profitability, and growth. Alternative growth strategies are increasing market penetration, developing new markets (types of customers or geographical expansion), acquisition of competitors, or

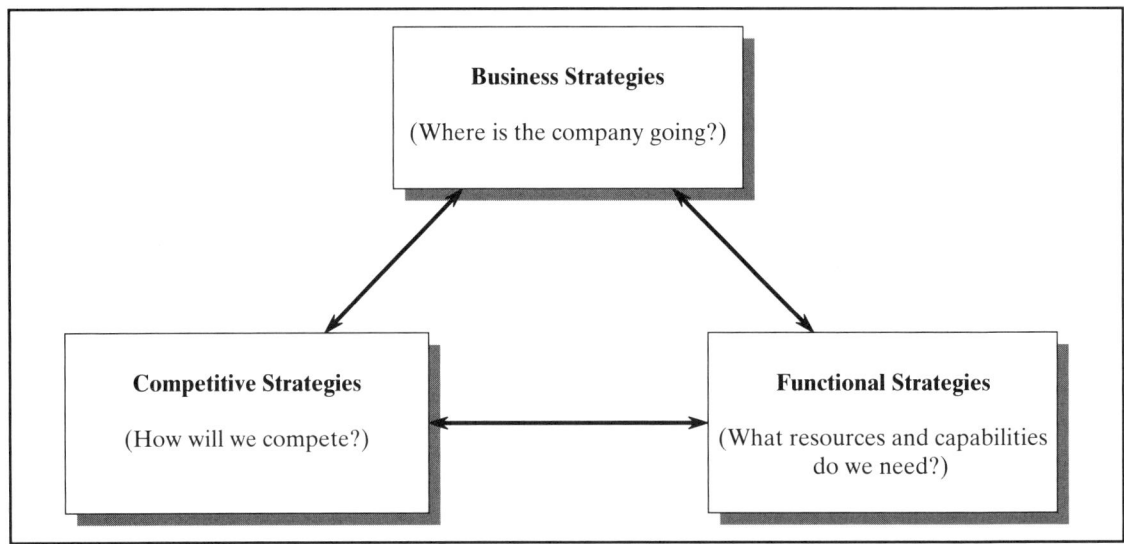

FIGURE 5.10 Levels of Strategies

diversification into other lines of business. The business strategies guide the development of the competitive and functional strategies.

The next set of strategies to be developed is the *competitive strategies*. They relate to creating or strengthening the firm's competitive advantage in the market. Competitive strategies address specific characteristics of the firm's services, such as price, quality, time to delivery, and post-construction services. They are selected to support in the accomplishment of the business strategies discussed above.

The last set of strategies to be developed is the *functional strategies*. These strategies provide the resources, capabilities, and core competencies needed to accomplish the business and competitive strategies. Typical functional strategies include the following:

- Service delivery strategies (capacity, location, process design, and management)
- Business development strategies (sectors, customers, and location)
- Human resource management strategies (recruiting and development)
- Research and development strategies (innovation)
- Information management strategies (system technology and information collected)
- Financial/accounting strategies (capital asset acquisition, investment, cash management, and debt management)

Let's look at an example. Suppose a construction firm has selected as one of its strategic objectives—*Increase company profitability by 3 percent per year over the next ten years.* The strategies selected for accomplishing this objective may include the following:

Business Strategies

- Increase the number of customers by 5 percent per year by expanding services and implementing new marketing initiatives.
- Control cost of doing business to increase profitability.
- Enter the design-build market.

Competitive Strategies

- Expand scope of service to include design-build by establishing partnerships with local design firms.
- Develop new marketing brochure to target new customers.
- Enhance recognition of company name by sponsoring charity events.

Functional Strategies

- Establish computerized project cost control systems to reduce cost of doing business.
- Establish a real-time overhead cost management system to control overhead costs.
- Recruit and develop additional staff to manage increased workload.
- Train project managers to manage design-build projects.
- Establish training program to train project managers and superintendents in presentation skills.

STRATEGY IMPLEMENTATION

Once the company's long-range strategies have been determined, short-term goals are identified and action plans are developed to implement the strategies. These short-term goals should be specific, measurable, result-oriented, and time-bound to guide the development of action plans. The planning horizon for these goals is one to two years. Suppose a construction firm has decided to grow its business volume from $50 million per year to $100 million per year over the next ten years. A two-year goal might be to increase the annual volume of work to $60 million per year. Short-term goals are selected for each of the strategic objectives previously identified.

Using the business, competitive, and functional strategies, action plans are developed to assign responsibility for goal accomplishment and to allocate resources. Each action plan identifies the goal to be accomplished, who has responsibility for its accomplishment, what resources (both people and financial) are assigned, and when the goal is to be accomplished. Forms similar to the example shown in Figure 5.11 may be used for developing action plans. The resulting strategic plan contains a series of action plans that assign responsibility and allocate resources. Because the future business environment may not evolve as forecast, a contingency plan should be developed to identify alternative strategies to adjust to the reduced demand for the firm's services, increased competition, or different government regulation.

STRATEGY EVALUATION

As was illustrated in Figure 5.1, strategic management is an iterative process that requires monitoring and periodic revision. Company performance may be reviewed monthly, but a quarterly review and analysis is more typical. Data relating to accomplishment of the firm's strategic objectives are collected and analyzed to determine whether or not progress is being made toward accomplishment of the short-term goals. The specific data collected may include contracts awarded, profitability, cost of doing business, customer satisfaction, marketing initiatives, employee recruitment,

Date: *January 15, 2008*			
Goal: *Increase business volume by 2% per year over next two years.*			
Assumptions: *Demand for construction will grow by 3% per year during the next two years, but contractor capacity will increase more rapidly so that the availability of construction services will continue to exceed customer demand.*			
Action	**Responsible Person**	**Resources Allocated**	**Date to be Accomplished**
Identify 25 potential new customers per year	Chief, Business Development Department	$40,000 per year plus one new staff member	January 15, 2010 (progress to be monitored quarterly)
Target the desired new customers by assigning project managers to cultivate relationships	Vice President for Construction Operations	$50,000 per year for cultivating relationships with new customers	January 15, 2010 (progress to be monitored quarterly)
Enhance relationships with existing customers to obtain additional work	Vice President for Construction Operations	$85,000 per year to enhance relationships with current customers	January 15, 2010 (progress to be monitored quarterly)

FIGURE 5.11 Sample Action Plan

absenteeism, and employee losses. For example, data that would be collected quarterly and evaluated for the action plan shown in Figure 5.11 would include

- value of work acquired during the quarter,
- number of prospective customers identified,
- prospective customer contacts by project managers, and
- current customer contacts made by project managers.

The data are reviewed by company executives to assess the progress related to accomplishing the company goals and objectives. If some functional area is behind, it may be that insufficient resources were allocated or that the business environment did not evolve as forecast. Planning assumptions may need to be reexamined, and goals or action plans modified. Annually, the entire plan should be assessed, and adjustments made where appropriate. Generally, the mission statement and vision do not need to be revised, but the external forecast and the internal conditions need to be reassessed. Strategic objectives may need to be modified, and alternative strategies selected. New short-term goals and action plans also may be needed.

Summary

Strategic planning involves assessing the company's strengths and weaknesses, analyzing the external environment, forecasting the future, and selecting objectives and action plans that provide corporate focus and guide company leadership.

Strategic management is an iterative process that involves monitoring and evaluating the external environment, measuring company performance relative to the accomplishment of its objectives, and reallocating resources, if necessary, to improve company performance. The initial step in developing a strategic plan is to develop a meaningful mission statement that explains why the company exists and what makes it different from other firms in the industry. Then a vision statement is crafted to describe where the company intends to be at the end of the planning period. Before selecting strategic objectives, a strategic assessment is conducted to develop a forecast of the future business environment and to assess the company strengths and weaknesses. Based on the forecast opportunities and threats, the internal assessment, and the company vision, measurable strategic objectives are selected to guide the company in accomplishing its vision. Strategies are selected to describe the methods to be used to accomplish the objective. Short-term goals are selected as intermediate targets while moving toward accomplishment of the strategic objectives. Action plans are developed to assign responsibility and provide resources for goal accomplishment. A feedback mechanism is established to collect data and periodically review company performance.

Review Questions

1. What is the purpose of the mission statement and why is it important to the strategic planning process?
2. What is the purpose of the company vision and why is it important to the strategic planning process?
3. What is the purpose of the strategic assessment?
4. How would you conduct an external analysis for a construction firm? Why is it an important part of the strategic planning process?
5. What factors would you consider in assessing the general business environment for a construction company?
6. What factors would you consider in assessing the specific industry environment for a construction company?
7. What factors would you consider in preparing a customer analysis for a construction company?
8. How would you conduct an internal assessment using a value chain analysis approach?
9. How would you conduct an internal assessment using an internal audit approach?
10. How would you conduct an internal assessment using an internal functional analysis approach?
11. What are the five topics that might be addressed by a construction firm's strategic objectives?
12. What topics do business strategies typically address?
13. What topics do competitive strategies typically address?
14. What topics do functional strategies typically address?
15. What topics are addressed in company action plans?
16. What type of feedback structure should be developed for the evaluation of company performance?

Exercises

1. You are the president of a general construction company that is currently doing about $10 million in business per year. One of the company's strategic objectives is to increase the volume of work by 50 percent in the next ten years. What specific strategies do you suggest the company adopt to achieve this objective?
2. You are the president of a construction company that specializes in the construction of shopping centers. What is your assessment of the general business environment in your local area for this type of company?
3. One of the strategies in your firm's strategic plan is to identify a suitable design firm with whom you could establish a joint venture and then pursue design-build projects. Write an action plan for implementing this strategy.
4. Prepare a strategic assessment of Cascade Builders using the data shown in Appendix B and develop a strategic plan for the company.
5. Prepare a strategic assessment of Northwest Constructors using the data shown in Appendix C and develop a strategic plan for the company.

Sources of Additional Information

Chinowsky, Paul S. *Strategic Corporate Management for Engineering*, New York: Oxford University Press, Inc., 2000.

Coulter, Mary K. *Strategic Management in Action*, 4th ed., Upper Saddle River, N.J.: Prentice-Hall, Inc., 2008.

David, Fred. *Strategic Management: Concepts and Cases*, 11th ed., Upper Saddle River, N.J.: Prentice-Hall, Inc., 2007.

Dess, Gregory, G. T. Lumpkin, and Alan Eisner. *Strategic Management: Creating Competitive Advantages*, 4th ed., New York: McGraw-Hill, Inc., 2008.

Friedman, Warren. *Construction Marketing and Strategic Planning*, New York: McGraw-Hill, Inc., 1984.

Porter, Michael E. *Competitive Advantage: Creating and Sustaining Superior Performance*, New York: The Free Press, 1985.

Wheelen, Thomas L. and J. David Hunger. *Strategic Management and Business Policy*, 11th ed., Upper Saddle River, N.J.: Prentice-Hall, Inc., 2008.

CHAPTER 6

Business Development

INTRODUCTION

Business development is the process of acquiring business for a construction company. This means retaining those customers the company wishes to retain as well as acquiring new customers with whom the company desires to do business. The basic components of business development are marketing and sales.

Marketing is the process of retaining existing customers, identifying prospective new ones, and attracting new customers to consider the construction firm as their service provider. Typical marketing functions include market research, public relations, and advertising. Marketing activities occur prior to a sales contact and include generating leads as well as networking to identify and screen prospective customers by learning their needs and attitudes. The objective is to attract customers to the company. It is about constant renewal of existing relationships and forging new ones.

Sales is contacting a specific potential customer identified by the marketing process and winning the construction contract. Typical sales functions include making presentations, submitting bids or proposals, negotiating, and executing contracts. Sales activities are customer focused and occur just prior to, during, and after a sales contact.

The business development process for construction services is shown in Figure 6.1.

Most construction firms are marketing a service rather than a product, because the construction site has been selected by the project owner, and the design has been completed by a design firm. Residential construction companies, however, often purchase a parcel of land, construct the utility and street infrastructure, construct homes, and then market the homes, which are products. To be successful, these companies must understand the product features desired by customers. Model homes are often used as marketing tools to show prospective customers product features.

Commercial, industrial, and heavy construction companies, however, are marketing a service. Unlike residential construction firms, these construction companies have no samples other than current and past projects and must select marketing tools that establish trust among prospective customers. This requires an understanding of the service characteristics desired by their customers. These characteristics often relate to cost, quality, service, and timeliness.

CHAPTER 6 Business Development

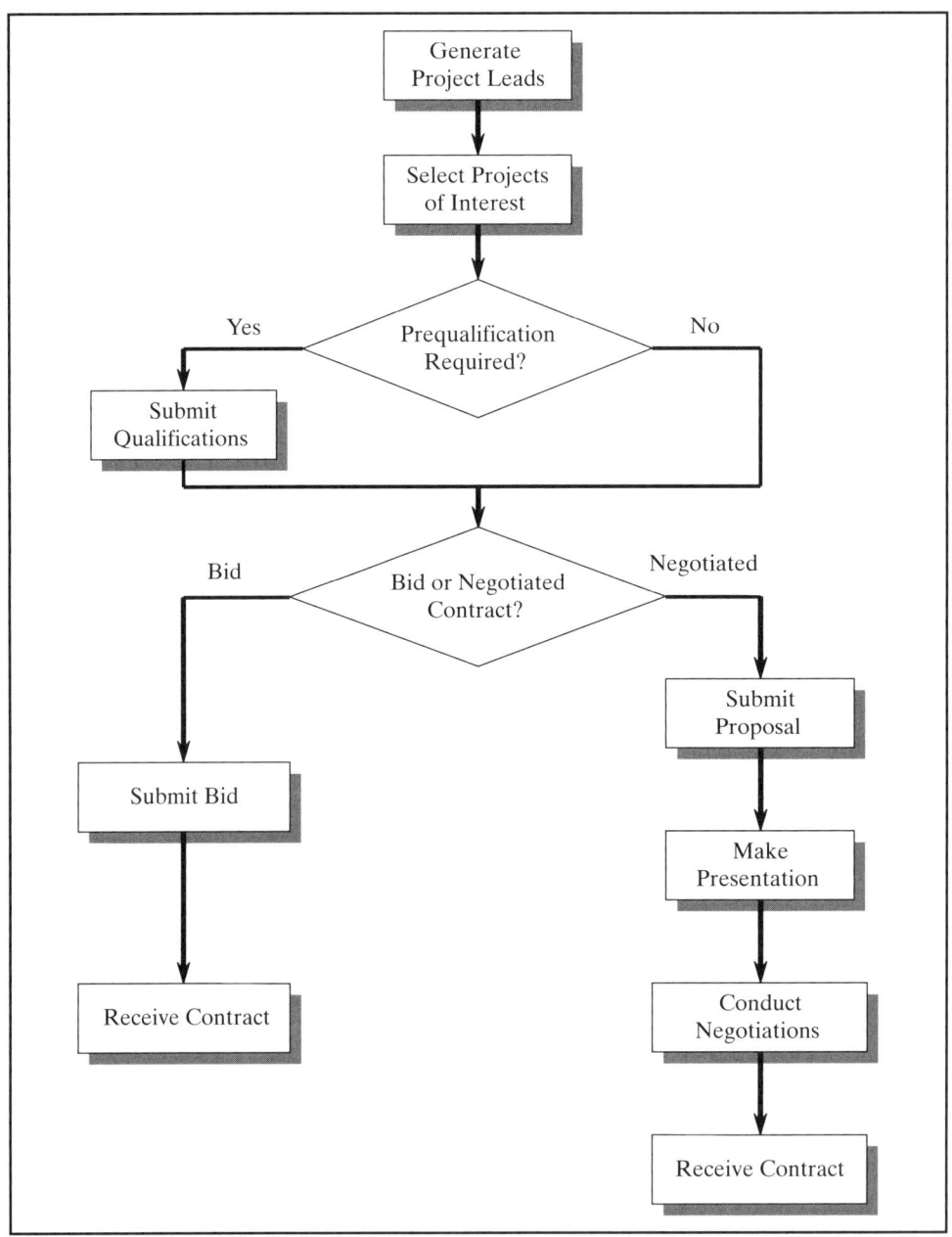

FIGURE 6.1 Business Development Process for Construction Services

The specific customers of interest were identified in the strategic planning process discussed in Chapter 5. A marketing plan is a supporting functional plan of the overall company business plan. Marketing strategies are selected that support the accomplishment of the company business strategies identified in the planning process. The strategic plan identifies

- the type of construction services to be offered by the firm (for example, construction only, design-build, or turnkey),
- its geographic service area, the size of projects to be undertaken (such as $10 million to $25 million), and
- the types of projects and customers (public, private, or both) to be pursued.

Historically, marketing was performed on a part-time basis by company owners and senior executives. Marketing activities were generally limited to brochures and advertising. This has changed since the 1990s, as most medium and large construction firms have established professional marketing staffs. The marketing staff tends to be an information gatherer providing information to company executives, project managers, and superintendents who actually make the sale. Most successful construction companies recognize that customers purchase services from people, not firms, and that relationship building generally is critical to making sales. Thus the people who deliver projects are often assigned the responsibility for cultivating and nurturing relationships with customers, subcontractors, and suppliers.

In this chapter, we will discuss the basic steps in developing a marketing plan. First we will examine the marketing process and some concepts for analyzing the company's market, which is similar to the external assessment discussed in Chapter 5. Then we will examine the development of marketing strategies to target prospective customers and some of the tools that can be used to accomplish the marketing objectives. Finally we will discuss the development of a marketing plan and the acquisition of work.

MARKETING CONSTRUCTION SERVICES

The basic function of marketing is to create business opportunities for the company. It involves everything that the company does, from the appearance of its project sites and its reputation for safe work to its company logo and the quality of its business correspondence. Marketing is an ongoing process. There is neither a beginning nor an end. Strategic marketing activities are undertaken by the firm to enhance its name recognition and reputation within the industry. Tactical marketing activities are used to target specific current or prospective customers. Internal marketing activities are used to keep the employees motivated to provide quality customer service.

The most important marketing resources are the firm's reputation and its satisfied customers. Success depends on a firm's ability to develop enduring relationships with targeted customers by continually meeting or exceeding their expectations and needs. This requires an understanding of the process by which potential customers are drawn to specific construction firms and become repeat customers and an understanding of the customers' specific project requirements. Retaining customers requires every employee to understand his or her responsibilities toward achieving customer satisfaction, because customer service is the most critical element of success in the construction business. Internal marketing is needed to ensure that the company's employees understand market and customer demands, their individual roles in achieving customer satisfaction, and their roles in achieving the company's strategic goals and objectives.

Marketing in the construction industry is quite different from marketing in many other industries. As stated in the Introduction, most construction companies are marketing a service, not a product. Generally, the product is described by a set of construction plans and specifications, or in the case of design-build, a set of design criteria. Construction firms market their services by emphasizing their ability to meet or exceed customers' criteria in terms of construction cost, time, quality, and safety. To many customers, achieving a specific project delivery date is more important than a minimum-cost project.

In most construction markets, the demand for construction services is cyclical with periods of strong demand followed by periods of weak demand. To be successful during periods of weak demand, construction firms must be resourceful, innovative, and focused on relationship building. They need to develop good reputations for quality work, reasonable pricing, timely delivery, and good customer service. Marketing initiatives need to emphasize the company strengths, or core competencies, in targeting potential customers.

MARKETING PROCESS

The basic steps in the marketing process are shown in Figure 6.2. The first step is to select the specific sector or sectors of the overall construction market that are of interest to the construction firm. This is known as the **relevant market**. This decision was made during the strategic planning process described in Chapter 5. The mission statement identifies why the company exists and what type of business it performs. The selection of customer types was made in order to perform the external analysis needed to develop the strategic plan. Once the relevant market has been defined, it is analyzed to determine which groups of customers have needs that best match the company's services and the selection criteria used by these prospective customers in selecting specific construction firms to construct their projects. Then the company's services are tailored to take advantage of its core competencies and deliver services that meet or exceed customer expectations.

The basic marketing concept is often denoted by the letters *STP*. The *S* means *segmenting* the market to determine the specific sectors of interest, *T* means *targeting* the specific set of customers of interest, and *P* means *positioning* the company for competitive advantage. Segmenting the market is somewhat like the work-breakdown structure used in developing a construction cost estimate. The market is segmented by general categories of construction, geographic area, type of owner, delivery method, and project size. This process is illustrated in Figure 6.3.

Each market segment of interest is analyzed to determine its current level of construction demand and its growth potential, the potential for projects being profitable, and the purchasing procedures used by prospective customers. Those customers whose needs and purchasing procedures best fit the company's core competencies and resources are targeted. The last step, positioning, is designing the company image, services, and marketing media so that the targeted prospective customers understand and appreciate the company's capabilities and customer service ethic in relation to those of its competitors. This requires an analysis of prospective customer needs and contractor selection criteria.

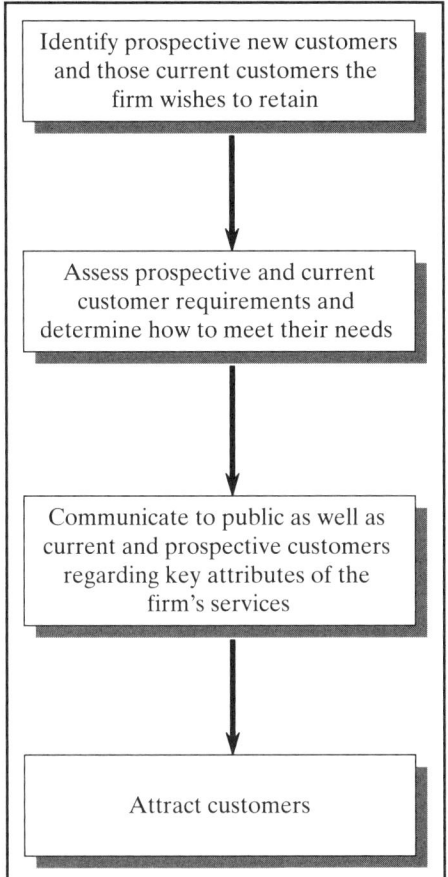

FIGURE 6.2 Marketing Process

Marketing strategies are selected that support the accomplishment of the company business strategies identified during strategic planning. These strategies identify the techniques to be used to market the company's capabilities to prospective customers. A specific marketing plan is developed to allocate resources and assign responsibilities. This is similar to the action plans discussed in Chapter 5. Marketing resources include

- Employees and their technical skills
- Equipment
- Financial position
- Reputation

Once the targeted customers have been identified, the construction firm needs to select the message to be sent to them and the media for transmitting the message. It could be a brochure, a Web site, a personal contact by an executive, by responding to a request for qualifications or proposal, or by requesting a referral from a current customer or a design firm.

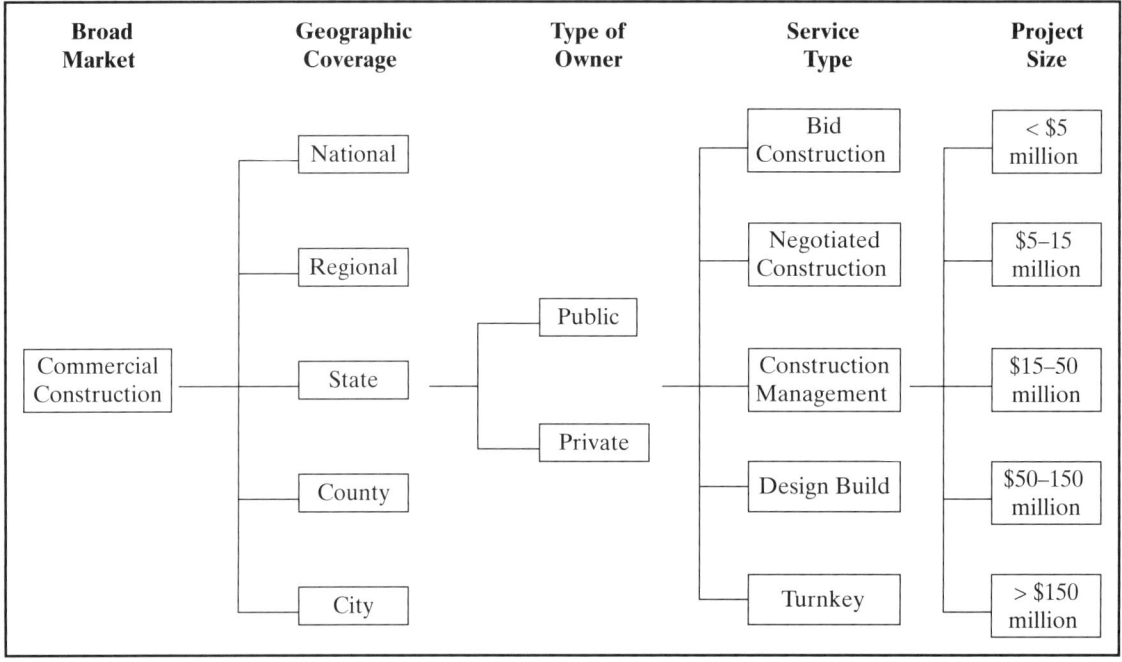

FIGURE 6.3 Segmenting the Market

In addition to seeking new customers, the construction firm needs strategies for retaining current customers with whom the firm desires repeat business. This is sometimes referred to as a customer relationship management plan. One method is to survey customers to determine their satisfaction with the company and its employees. Another is for senior managers to make periodic visits with current customers. Project managers are often assigned the responsibility for cultivating long-term relationships with past customers and determining their future plans. Time and funding must be given to the project managers if they are to be expected to perform marketing duties. Typically, a half-day per week should be sufficient.

MARKET ANALYSIS

During the development of its strategic plan, the construction firm determines the specific set of services to be offered and the segment or segments of the overall construction market that contain the customers of interest. For example, suppose Pacific Constructors decided to offer design-build, construction and construction management services to private customers who construct commercial projects with values between $5 million and $50 million in California and Arizona. This then defines six relevant markets that need to be analyzed, because customer selection criteria for construction services may be different from those interested in design-build or construction management services and the anticipated demand in California may

CHAPTER 6   Business Development   123

differ from that in Arizona. There are three components to the market analysis—assessment of current and projected customer demand, assessment of customer satisfaction, and assessment of current and potential competition.

Demand Assessment

The basic steps in assessing the demand for the services offered by the company are illustrated in Figure 6.4. The first step is to estimate the total demand within the geographic area and sector or sectors of interest. This is similar to the external analysis (opportunities and threats) discussed in Chapter 5. The United States Census Bureau

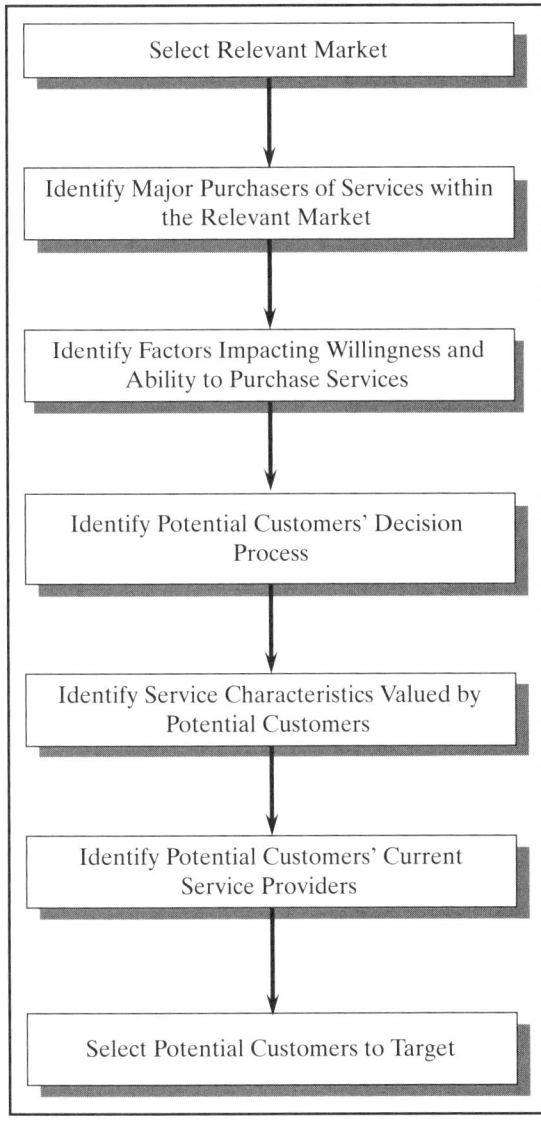

FIGURE 6.4 Demand Assessment Process

publishes a *Census of the Construction Industry* every five years, which indicates the historic volume of major types of construction performed in each state. This database can be used to determine the volume of private sector commercial construction in California and Arizona. Using this data and an economic forecast, the leaders of Pacific Constructors forecast the future demand for commercial construction. This total demand is then divided into the specific types of services customers may desire. The portions of the total demand, or segments, relating to construction, design-build, and construction management are the segments of strategic interest to Pacific Constructors. These segments are further divided by project size to determine the relevant markets for the firm.

Once the relevant markets have been defined, the company identifies the potential customers in each market, their willingness to purchase the types of services offered by the firm, and their ability to purchase the services. Other considerations are

- What decision criteria do they use in the selection of prospective construction firms, and how do they select the winning firm?
- When considering prospective construction firms, do they consider any firm that is interested or do they limit their consideration to a few select firms?
- Do they have long-term relationships with competitors and not consider other firms?

Based on this analysis of the relevant markets, the construction firm selects the specific set of potential customers to target with its marketing efforts.

Customer Satisfaction Assessment

The key to retaining current customers is to ensure that they are satisfied with the services that they have received from the construction company. One approach is to make personal contact with decision makers in the customer firms to assess their satisfaction. Another is to survey each customer at the end of each project, using an evaluation form similar to the one shown in Figure 6.5.

While the company might not offer the least-cost service, the customer needs to perceive that added value was obtained for the additional cost. Customer value may be defined as

$$\text{Customer Value} = \frac{\text{Perceived Performance}}{\text{Delivered Cost}}$$

This is not a mathematical formula, but a way to assess value obtained per dollar of investment. For example, the owner can assess project performance on a scale of 1 to 5, with 5 being the highest. Suppose the owner has completed ten projects constructed by three construction firms. The owner can assess his or her satisfaction with each project and rate it from 1 to 5. Then using the relationship shown above, the owner can evaluate the customer value provided by each of the three construction firms and use this information when selecting a construction firm for a future project. If customers are not satisfied with the value of services being received, the construction company must improve some characteristics of its service delivery to obtain a more competitive advantage.

Customer Evaluation

Project: _____

Please rate our work and our company, so that we may improve our customer service.
Thank you for your time.

	Excellent	Average			Poor
	5	4	3	2	1
Project Evaluation:					
Within Budget					
On Time					
Quality of Work					
Responsiveness					
Safety					
Company Evaluation:					
Customer Communications					
Internal Communications					
Teamwork					
Service Attitude					
Technical Skills					

Comments: _____

FIGURE 6.5 Customer Evaluation Form

Competition Assessment

Before selecting marketing strategies, a construction company needs to assess its competition in each of its relevant markets. The company needs to identify existing competitors, assess the likelihood of new entrants, and evaluate the services currently offered, or likely to be offered, by competitors. A worksheet similar to the one illustrated in Figure 6.6 can be used to analyze each competitor. In most instances, a construction firm will have less than ten firms in each relevant market that it considers its competitors. A worksheet is prepared for each one. The objective is to identify the competition and assess the value of the services they offer. Understanding competitors' services is essential if the construction firm is to tailor its services and marketing message to provide competitive advantage. It is also suggested that the marketing staff keep track of projects received by each of the company's primary competitors and that an assessment be made to determine why the competitor was selected for each project.

Competitor Evaluation

For each major competitor, answer the following questions:

1. Name and location of competitor (main office): _____

2. Estimated size of staff: _____

3. Types of customers targeted by competitor: _____

4. Marketing techniques used by competitor: _____

5. Major strengths of competitor: _____

6. Major weaknesses of competitor: _____

7. Primary reasons why customers should buy our services rather than competitor's:
_____

FIGURE 6.6 Competitor Evaluation Worksheet

MARKETING STRATEGIES

Marketing objectives are identified in the strategic plan. They typically relate to market share or sales volume. Strategies are then selected to achieve these objectives. Marketing strategies are developed at three levels—strategic marketing, tactical marketing, and internal marketing.

- **Strategic marketing** strategies are those selected to enhance company name recognition and its reputation. These might involve changing the company logo, sponsoring a youth sports team, advertising in trade publications, participating in trade shows, or repainting company vehicles. Image builders include

- Clean jobsites and company facilities
- Neatly attired employees
- Courteous, professional behavior from all employees
- Promptness for meetings and service calls

- **Tactical marketing** strategies are those selected to target specific current or prospective customers. These might include targeted mailing of focused brochures, cold calls by project managers, and sponsored social activities.
- **Internal marketing** strategies are those selected to inform employees of company activities and ensure that they understand their role in maintaining the company's reputation and in marketing its services. This might include newsletters and training sessions. It is also important to keep employees motivated by publicly recognizing their service and accomplishments.

There are three alternative tactical marketing strategies that might be adopted to increase sales. One is to increase the demand for the type of services offered by the company. This is almost impossible to accomplish, because construction firms generally have little influence on customers' desire or ability to purchase construction services. It is possible, however, to increase the demand for a specific service by convincing a customer to select a different delivery method. For example, using a design-build approach rather than using the traditional design-then-build approach.

Another tactical strategy is to move existing customers up the "buying cycle," which is illustrated in Figure 6.7.

FIGURE 6.7 Buying Cycle

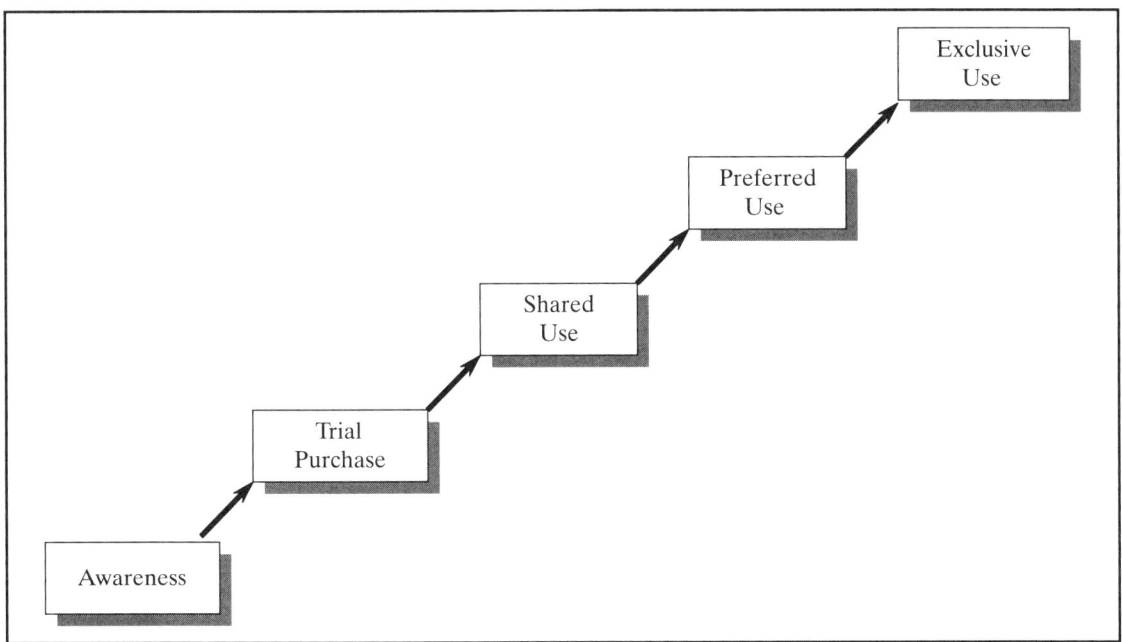

- *Awareness* means the prospective customer is aware of the construction firm and its service capabilities.
- *Trial purchase* means the customer has decided to select the firm for a single project.
- *Shared use* means the customer has decided to use the construction firm and one or two of its competitors for all projects.
- *Preferred use* means the customer uses the construction firm for most of its projects, but occasionally uses other construction companies.
- *Exclusive use* means the customer has decided to use the construction firm for all its projects.

Moving customers up the "buying cycle" requires the development of enduring close personal relationships at all levels between the construction firm and the customer's organization.

The third tactical marketing strategy is to acquire new customers by attracting them away from the firm's competitors. This may mean entering new markets or diversifying and offering additional or altered services. To be successful, marketing tools are needed that demonstrate the better value provided by the construction company's services in relation to those offered by its competitors.

One of the major decisions to be made in the selection of marketing strategies is the decision regarding how to price the services the company offers. The key factors to be considered in this decision are illustrated in Figure 6.8. Services must be priced above the cost of delivering them if the firm is to generate a profit. Services may be priced above those of competitors if service attributes provide more value for the customer's investment.

The central issue in marketing is the development of a communications strategy. This involves answering the following questions:

- What is the message?
- Who is to read the message?
- How is the message to be communicated?
- What effect on the reader is desired?

Some communications will be targeted at the general public, others at current customers and prospective customers, and still others internally at the company's employees. Each communication tool should be narrowly focused at the target audience to achieve the maximum effect.

FIGURE 6.8 Service Pricing Factors

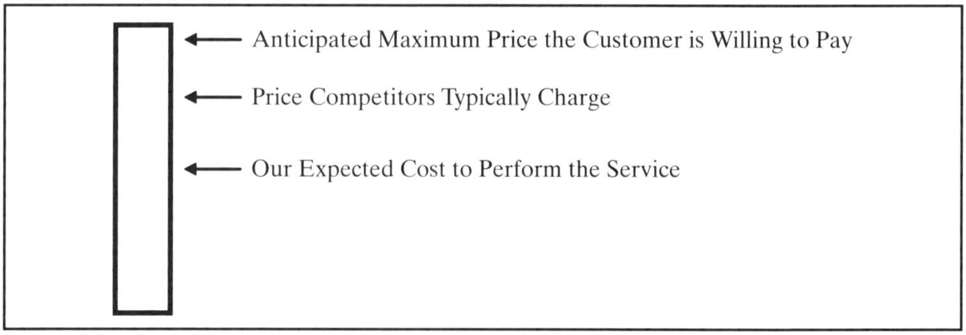

- Anticipated Maximum Price the Customer is Willing to Pay
- Price Competitors Typically Charge
- Our Expected Cost to Perform the Service

Marketing is not free. It requires both human and financial resources and is a cost of doing business. Many construction firms spend 1 to 3 percent of their annual gross sales on business development. A marketing plan is used to identify the specific marketing objectives for the firm, the strategies to be used in accomplishing these objectives, and the action plans that assign responsibilities and allocate resources.

MARKETING TOOLS

There are many tools that may be used to support the company's marketing activities. The major ones are discussed in this section. The potential use of these tools to accomplish strategic, tactical, or internal marketing goals is shown in Figure 6.9.

Company Reputation is the most important marketing tool. Everything a company does goes into creating the company's reputation. Without a good reputation, a company will have difficulty increasing its business volume.

Company Logos are used to enhance the public image of a construction firm. A company's logo should become an integral part of every visual marketing communication, including company stationery and business cards. Logo colors should be selected to differentiate the company from its competitors. The design

FIGURE 6.9 Uses of Marketing Tools

Marketing Tool	Strategic Marketing	Tactical Marketing	Internal Marketing
Company Reputation	✓	✓	
Company Logos	✓	✓	
Business Cards	✓	✓	
Stationery and Business Forms	✓	✓	
Brochures	✓	✓	
Newsletters	✓	✓	✓
Advertising	✓	✓	
Market Reference Files	✓	✓	
Networking	✓	✓	
Public Relations	✓		
Web Sites	✓	✓	✓
Subcontractor Relationships	✓	✓	
Supplier Relationships	✓	✓	
Statements of Qualification		✓	

should be simple and have good visibility. Because the logo will be linked to the firm's public image, a graphic designer should be hired to create it. The logo should look good both in black and white and in color. Clean, spare designs are better than complex ones. The logo should be used on stationery and business forms, business cards, hard hats, project signs, and vehicles.

Business Cards are used to identify employees with the company. All cards should have a similar appearance with the company logo and the name, address, telephone number, fax number, and e-mail address of the individual. Some companies print their mission statement on the back of all business cards. Use of attractive business cards helps develop the professional reputation of the construction firm.

Stationery and Business Forms are used both for communication and for marketing. Professionally prepared stationery and business forms that bear the company's logo help enhance the company's image as a professional organization. All correspondence needs to be proofread carefully to further enhance the firm's professional reputation.

Brochures are used to create initial impressions among prospective customers. They should be developed with sufficient flexibility to allow for tailoring to target specific categories of prospective customers. Brochures should not be allowed to become stale, but should be replaced about every two years. The brochures should be structured to get the reader's attention and to communicate the company' philosophy, core competencies, and record of successes. Grab the readers' attention with the cover and tell them who you are and what you do. Brochures should be simple, but focused to target prospective customers.

Newsletters are used to communicate successes to current and former customers, to sustain relationships and to publicize employee successes. They are also used to communicate with employees to inform them of new customers, projects, and company initiatives and to recognize employee accomplishments. They should be published regularly in a succinct, consistent format.

Advertising is used to publicize the company, its capabilities, and its service offerings. Advertising includes direct mail, yellow pages of telephone books, advertisements in trade publications, displays at trade shows, project signage, company logo on vehicles, and give-away items, such as pens, cups, hats, and jackets. Advertising is used to increase the public and prospective customer awareness of the company and to gain greater recognition for its name and capabilities.

Marketing Reference Files are used to collect information regarding current and prospective customers and competitors. They should contain information about projects that competitors acquired and perceptions as to why they were successful. Reference files should also contain information regarding anticipated projects that current and prospective customers might undertake.

Networking is participating in professional association activities as well as community and charity events. The goal is to make personal contacts with prospective customers and designers and to enhance the professional image of the construction company.

Public Relations are initiatives to enhance the public image of the company. They may include sponsoring youth sport teams and public service and charity events. Senior company leaders may participate in community service organizations, such as Rotary and United Way, to enhance the image of the company within the community. Another public relations technique is to submit short articles about company projects to local newspapers and trade publications and to invite reporters to visit project sites. News releases should be submitted to local media announcing new projects, new employee hires, employee promotions, and completed projects. Reprints of these articles can be used to support marketing efforts.

Web Sites are used to provide information about the company, its capabilities, and its current projects. Some construction firms have cameras linked to their Web sites so that viewers can see the current condition of major construction projects.

Subcontractor Relationships are not only important for the construction of quality projects, but they can also aid marketing initiatives. Good subcontractor relationships strengthen the construction firm's reputation with customers and may help in securing new projects.

Supplier Relationships are also important to the construction firm's reputation. The ability to obtain quality materials at reasonable cost is important to the firm's ability to provide customer value.

Statements of Qualifications are critical to the acquisition of negotiated construction contracts. Project owners typically issue a *Request for Qualification* to interested construction companies. Based on evaluation of the *Statements of Qualification*, the project owner selects the finalists for an interview and a submission of proposals. Good *Statements of Qualification* should be project-specific and focus on the type of project that the owner intends to construct.

MARKETING PLAN

Marketing without a plan is like attempting to construct a building without a set of drawings. The process for developing a marketing plan is illustrated in Figure 6.10. The marketing plan describes the marketing objectives, the strategies, and the action plans the company will use to accomplish its overall company objectives. It is structured to support the accomplishment of the company strategic objectives identified during the strategic planning process. The plan needs to address the following phases of customer development:

- Obtaining access to the customer
- Orienting the customer to the services provided by the construction company
- Cultivating a relationship with the customer
- Maintaining the customer relationship

For example, suppose a construction company established a strategic objective to increase its business volume by 3 percent per year over the next ten years. To accomplish this objective, the company may need to broaden its management systems and to recruit and develop additional project management teams. It also needs to develop marketing strategies to develop additional business opportunities. One set of strategies should be focused on maintaining relationships with existing customers. This may

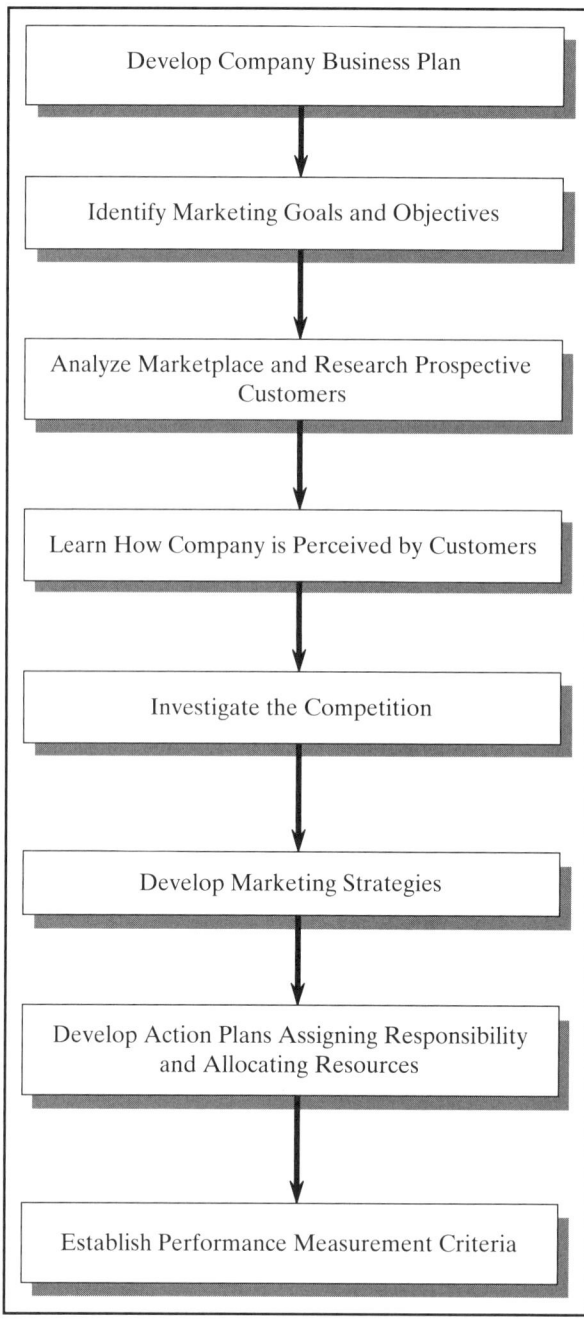

FIGURE 6.10 Marketing Plan Development Process

include providing them with newsletters and tasking project managers to make monthly contacts with their counterparts in customer organizations. Another set of strategies is needed to identify and develop new customers. This may involve publishing a new brochure, advertising, networking, participating in trade shows, and making cold calls on targeted organizations. A third set of strategies is needed to enhance the construction

	Health Care	**Hospitality**	**Higher Education**	**Commercial Office**
Project Type	Hospitals, clinics, and laboratories	Luxury hotels, casinos, and resorts	Community colleges and universities	Mid- to High-Rise
Geographic Area	Western U.S.	Northwest U.S.	Western U.S.	California
Min/Max Size	$20–$100M	$40–$200M	$10–$75M	$15–$125M
Volume	$150M	$150M	$100M	$100M
Fee	$4.5M	$6.0M	$3.5M	$4.0M

FIGURE 6.11 Marketing Plan Overview

firm's public image and reputation. This may involve sponsoring charity events or youth sports activities. Once the specific strategies have been selected, specific action plans are prepared allocating resources, assigning responsibility, and establishing time frames for accomplishment. A fourth set of strategies is needed for internal marketing. Newsletters and training events or workshops are often used for this purpose.

Once the market analysis has been completed, it is often useful to quantify, for the company, the specific marketing goals for each market sector in which the company aspires to compete. An example is shown in Figure 6.11. These goals are then used to craft marketing strategies for their accomplishment. These quantifiable goals can then be used to measure the progress in accomplishing the company's overall business plan.

ACQUISITION OF WORK

Once the target customers have been identified, it is time for the sales team to get into action. If the prospective customer selects construction firms solely on price, such as many public owners, the sales team must prepare a responsive bid based on the contract documents. If the owner uses an interview and proposal process, the sales team must develop a winning presentation that can engage the customer followed by a winning proposal that fully responds to the customer's *Request for Proposal*. The structures of both the interview presentation and the proposal must be based on the project owner/s values, interests, and project objectives.

Summary

Business development is the process of acquiring business for the company. It means retaining those customers the company wishes to retain and attracting others with whom the company desires to do business. The basic components are marketing and sales. Marketing involves retaining existing customers and attracting new ones. Sales is winning a construction contract. Market plans are developed to support the accomplishment of the corporate objectives identified in the strategic plan. The most important marketing resources are the company's reputation and its satisfied customers. The basic marketing concept is to segment the market, to target customers of interest, and to position the company for competitive advantage. Market analysis involves assessing the current and potential customer demand, customer satisfaction, and potential competition. Based on the market analysis and overall company strategic objectives, marketing strategies are selected. Strategic marketing strategies are identified to enhance the company's reputation. Tactical marketing strategies are selected to target specific customers. Internal marketing strategies are developed to keep company employees informed. Specific marketing tools are selected to communicate the message to the various audiences of interest. Action plans are prepared to allocate resources, assign responsibility, and assign time frames for action accomplishment.

Review Questions

1. What is the difference between marketing and sales?
2. What are the most important marketing resources for a construction company?
3. What are the two primary goals of marketing?
4. What is meant by the term "relevant market?" How does a construction firm determine its relevant market?
5. Explain the marketing concept denoted by the letters STP.
6. What is a demand assessment and how is it conducted?
7. How would you assess the satisfaction of current customers?
8. What are strategic marketing strategies selected to achieve?
9. What are tactical marketing strategies selected to achieve?
10. What are internal marketing strategies selected to achieve?
11. Explain what is meant by moving a customer "up the buying cycle."
12. What factors should a construction firm consider when determining how to price its services?
13. What factors would you consider in selecting a logo for a construction company?
14. Why do construction companies develop brochures? What basic information do they contain?
15. Which marketing tools would you use to identify and develop new customers?

Exercises

1. You are preparing a marketing plan for your company. One of its strategic objectives is to increase its volume of work by 2 percent per year over the next ten years. How would you develop a marketing plan to support the accomplishment of this strategic objective? Which marketing tools would you use?
2. Your construction company has historically done only public sector work. Because of a decline in government spending, it has decided to pursue privately funded projects. What type of market analysis should the company perform?

3. Your construction firm has historically done only mechanical construction business in California. You have decided to expand and offer design-build mechanical construction services in Nevada. What information would you gather in developing a marketing plan for this geographic and project delivery expansion?
4. Develop a business development plan for Cascade Builders using the data shown in Appendix B.
5. Develop a business development plan for Northwest Constructors using the data shown in Appendix C.

Sources of Additional Information

Armstrong, Gary. *Principles of Marketing*, 12th ed., Upper Saddle River, N.J.: Prentice-Hall, Inc., 2008.

Friedman, Warren. *Construction Marketing and Strategic Planning*, New York: McGraw-Hill, Inc., 1984.

Goncalves, Karen P. *Services Marketing: A Strategic Approach*, Upper Saddle River, N.J.: Prentice-Hall, Inc., 1998.

Hiebing, Roman G. and Scott W. Cooper. *The Successful Marketing Plan: A Disciplines and Comprehensive Approach*, 3rd ed., New York: McGraw-Hill, Inc., 2003.

Kubal, Michael T., Kevin T. Miller, and Ronald D. Worth. *Building Profits in the Construction Industry*, New York: McGraw-Hill, Inc., 2000.

Palmer, Adrian and Catherine Cole. *Services Marketing: Principles and Practices*, Englewood Cliffs, N.J.: Prentice-Hall, Inc., 1995.

Pickar, Roger. *A Contractor's Guide to Focus Sales and Increase Profitability*, Washington, D.C.: Associated General Contractors of America, 1995.

Society for Marketing Professional Services. *Marketing Handbook for the Design & Construction Professional*, Los Angles, Calif.: BNi Publications, Inc., 2000.

CHAPTER 7

Human Resources Management

INTRODUCTION

Human resources represent the social capital of a construction firm and should be managed as carefully as financial assets and capital investments. The company's core competencies reside in its employees, and it is they who provide the customer satisfaction that is needed to retain customers and to remain a viable business enterprise. Adequate skilled and motivated human resources are essential if a company is to improve its business performance and develop an organizational culture that fosters innovation. While company leaders set the goals for the firm, it is the employees who take action to achieve them.

Company leaders have a significant role in influencing their employees' perceptions of and satisfaction with the company. The ability of a construction firm to achieve and maintain success is largely dependent on management decisions; the company culture; the company's ability to attract and develop loyal, motivated employees; and the company's ability to retain employees.

Human resources management programs must be linked to the strategic planning process to reduce any weaknesses identified and to attract and to develop the workforce needed to be able to successfully handle the challenges forecast for the future. Investing in human resources development is essential to remaining a competitive construction firm.

In this chapter we will discuss the major issues associated with human resources management. These are company culture, organizational design, staffing, employee development, performance management, employee rewards and benefits, employee retention, union relations, safety and wellness, and the regulatory environment. To be most effective, the company's human resources management programs, policies, and practices should be structured to support the mission, the vision, and the company objectives contained in the strategic plan. Investment in human resources development should be a key element of the company's strategic plan.

THE CHALLENGE

There are two major challenges relating to human resources management that company leaders face. The first is to attract employees, develop them, and retain them. This is primarily a function of how employees are treated within the company. Meeting this challenge is becoming more difficult as current employees retire, and talented replacements are hard to find. The second challenge is to ensure that company human resources management policies and procedures conform to legal requirements. Suggested strategies to meet both of these challenges are addressed in this chapter.

The chief executive officer or owner of a construction company is the focal point that everyone within the company looks to for direction, for leadership, for values, for recognition, and for assurance that everyone within the company is moving in the correct direction. Company leaders, knowingly or not, form their companies in their own image. They set the standards for company ethics and values. These leaders must understand that the success of their companies lies in their abilities to meet the needs and expectations of both their employees and their customers. Company leaders need to listen to their firms' employees and customers and then formulate goals, methods, and directions for their companies.

The effectiveness of the company leadership can be measured in the following three ways:

- The company's business results
- Its organizational structure
- Its culture

Company leaders affect the business results not only through changes in the organizational structure, but also in the work culture that they create. They establish the company culture by

- identifying and communicating core values,
- specifying expected employee behaviors,
- establishing methods for providing performance feedback to employees, and
- creating supporting recognition and reward systems.

As discussed in Chapter 1, company leaders must understand how people behave in organizations and how to motivate people to perform and want to remain as members of the company team. The primary measures of employee motivation identified are job performance, absenteeism, and turnover.

The quality of job performance is influenced by

- an employee's understanding of the importance of the job,
- the employee's perception of how he or she is valued within the company,
- the employee's compensation,
- whether or not he or she is publicly recognized for the quality of work performed, and
- the working environment.

Absenteeism is generally a function of an employee's

- dissatisfaction with the work environment,
- job responsibilities, or
- lack of recognition.

Turnover is influenced by the same factors as absenteeism as well as

- compensation,
- the availability of opportunities for development and advancement, and
- the perceived opportunities at another firm.

Construction companies are composed of unique individuals who possess different personalities and attitudes that can affect their behavior at work. Company leaders need to determine which positions best fit each current and prospective employee's personality. Individuals whose personalities are not compatible with the job requirements of a position might not be successful in that position.

The last aspect of organizational behavior to be considered is how people behave within an organization. Construction companies function best when members work together as teams to accomplish collective tasks. These may be formal teams, or they may be informal teams. The strength of the company comes from the synergy of employees working together to accomplish collective goals and objectives. Understanding how to motivate people to work together is a major challenge for company leaders.

DECISION-MAKING PROCESSES

While a company's leadership is a significant shaper of the contributions of the company's employees, the role that management decisions play in increasing or decreasing employee motivation often goes unrealized. Adopting a strategic planning process and a decentralized decision-making approach generally encourages employees to invest their time and effort in cultivating new business by strengthening existing customer relationships.

These actions align the values of the employees with those of the company. The result is an effective, efficient way to develop a lasting competitive advantage; namely, creating opportunities for employees to develop and succeed within the company. When employees are all working toward the same collective goals, the company is able to realize economic gains while managers are able to focus on developing business strategies for the future.

While compensation is increasingly cited as a much lower factor in overall employee satisfaction than is generally thought, management decisions regarding compensation strategies can be a significant issue if handled in a way that is perceived as unfair. Management decisions in the area of employee compensation strategies often serve to either reinforce or negate the importance of consistency. Significant horizontal pay differences contrive a perception of inequitable pay—inducing job dissatisfaction, low effort, and turnover.

Managers directly influence these perceptions by their ability to recognize all members of the team—those with more complex and highly compensated jobs as well as those whose responsibilities are more straightforward—as important to the success of the company. For all employees to feel valued and worthwhile, there must be a pervasive attitude that everyone's work is important. This attitude starts at the top and is reinforced daily by frontline supervisors.

COMPANY CULTURE

Almost every day managers and employees both face situations that have the potential for either furthering or hindering the success of a company. When handled poorly, these situations can turn into political battles, decreased productivity, angry clients, and a demoralized workforce. How these situations are handled can make the difference between average and superior results.

While every company has its own unique culture, most companies do not consciously try to create a certain culture. Typically, the culture of the organization is created unconsciously, based on the values and actions of the company leaders. The manifestation of top management's perceived values and actions is the establishment of normal practices for how demanding situations must be handled by employees. These normal practices in turn establish the corporate culture for a company. Company leaders' role in establishing ethical standards for their employees was discussed in Chapter 1.

Employees are constantly reacting to a set of expectations that they believe are being asked of them. This includes the way they relate to others. Seeing this, corporate culture should not be ignored or taken to be a static thing. Rather, it should be addressed in the company's mission, vision, and goal statements, and emphasized in company-sponsored training and in the communications employees receive from their managers.

Management's role in creating a positive company culture is to provide employees with opportunities to learn to be more effective as leaders and to maintain an increased awareness of the impact of their actions and decisions on others. As top managers gain experience in how to motivate, they inspire others to achieve immediate and sustainable results—the end result for the company being the development of talented people running major activities.

How decisions are made establishes a large part of a company's culture. When assessing the culture in effect at any given time, it is critical to bear this in mind because frequently, the root cause of failure is confusion regarding decision-making authority. The confusion often results in a power struggle. Like most power struggles, the end result is that someone loses, often the company through lost time and productivity.

The obligation to be accountable and responsible is enhanced within a culture that supports individual and collective empowerment. The conditions that help to create a culture of empowerment include

- The authority to act
- Adequate resources to act
- An environment of trust
- An acceptance of the obligation to be both accountable and responsible

To the extent that these conditions exist, employees will feel empowered to act individually and collectively. Individual managers then become personally accountable and responsible to their supervisors or team leaders for their own actions and for all aspects of their subordinates' activities.

A company with a positive culture empowers employees and rewards their successes while letting them learn from their mistakes. When important decisions have to be made, the culture will steer employees to tap everyone's best thinking, to work efficiently, and to get superior results. Employees will work to identify the real issues clearly and effectively and get unbiased views of what would be best for the company and/or project from other team members.

The company culture significantly influences the company's reputation as a desirable place to work. How people are valued and treated will affect whether or not they wish to be long-term employees. Most construction projects are accomplished by decentralized teams working together to meet customer expectations. Creating a collaborative team atmosphere is essential. Decision making should be decentralized to empower employees to recognize project execution issues and take action to resolve them.

ORGANIZATIONAL DESIGN

Organizational design involves the selection of the organizational structure for the company and the determination of the number of people needed for each organizational element. The alternative types of organizational structures were discussed in Chapter 2. It is extremely important to select an organizational structure that is compatible with the company culture, particularly at the decision-making levels. The output of this analysis is the organization chart for the company, which depicts the lines of supervision and authority.

Once the organization chart for the company has been developed, a job analysis is conducted for each position. Job analysis is gathering information about each job and is the primary building block for most human resources management systems, as illustrated in Figure 7.1. The major tasks, duties, and activities to be performed by the person occupying each position are identified. The task inventory includes determining

- the information input,
- the mental and/or physical processes to be performed,
- the work environment,
- the relationships to other jobs, and
- the characteristics of the work output.

Once the task inventory has been completed, the critical knowledge, abilities, skills, and other characteristics needed to be able to perform the job are identified.

The output of job analysis is a job description and a set of job specifications for each position. The **job description** defines the major tasks and duties to be performed

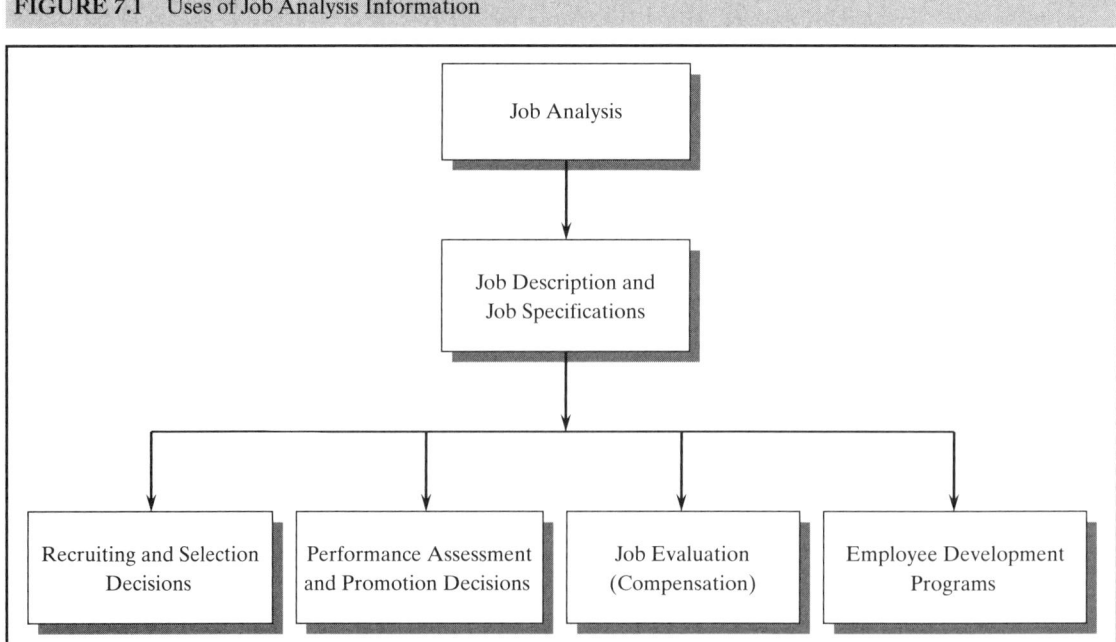

FIGURE 7.1 Uses of Job Analysis Information

by the person selected for the position. The **job specifications** identify the knowledge, the abilities, and the skills needed to be able to perform the tasks and duties contained in the job description. An example of a job description and a set of job specifications for an administrative assistant is shown in Figure 7.2. The information developed through the job analysis is used in

FIGURE 7.2 Job Description and Job Specifications for an Administrative Assistant

Job Title: Administrative Assistant

Reports to: Project Manager

General Responsibilities: Provides clerical and reception services for project field office. Assists project personnel in the preparation and processing of written, verbal, and electronic communications. Responsible for the operation and maintenance of project office equipment.

Specific Responsibilities:

1. Provides reception services for the project office. Receives, records, and directs site visitors, telephone communications, electronic mail, internal distribution, parcel deliveries, and postal mail.
2. Assists the project manager in maintaining project files. Copies project documents as required by the project manager and company operating procedures.
3. Provides data processing support to the project office. Inputs data as required into computer database management, spreadsheet, and other applications programs.
4. Prepares final copies of letters, memorandums, typed standard forms, and other written communications, using conventional typewriters and word processing programs.
5. Prepares time sheets for hourly employees.
6. Coordinates maintenance of field office and office equipment to maintain a presentable, efficient, and functional environment. Ensures that all critical office equipment are operational at all times. Responsible for maintaining an operational stock of consumable office supplies.

Required Knowledge, Skills, and Abilities:

1. Knowledge of construction terms and construction documents.
2. Knowledge of the operation and user-level maintenance of standard office equipment.
3. Knowledge of computer database management, word processing, and spreadsheet software applications.
4. Skilled in keyboard operations for fast and accurate typing and data input.
5. Able to communicate effectively in person and over the telephone. Able to maintain a focus on customer service and to represent the company at the project in a professional, courteous manner.
6. Able to organize, plan, and prioritize work efficiently. Must be able to work independently and productively with limited resources.

Experience Requirements: This position requires a minimum of five years' experience in office administrative support functions, with a minimum of two years' experience in construction field offices.

- recruiting and selecting new employees,
- promoting current employees,
- managing employee performance,
- setting compensation levels, and
- structuring employee development programs.

STAFFING

There are two major issues associated with attracting prospective new employees. The first is to attract them to the construction industry, and the second is to attract them to one's firm. Internet Web sites are an effective tool for advertising career opportunities in a specific company. The Web pages should clearly articulate the company's mission, vision, and values and identify the critical competencies for organizational success (knowledge, skills, and behaviors). Career development programs, types of job experiences, and total compensation packages available should also be discussed.

The key to recruiting is to emphasize the diverse type of work the firm does and describe the company as a rewarding place to develop a career. Recruiting at technical schools, community colleges, and universities is also effective. Some management personnel may be recruited from the ranks of craft labor, while placement firms may be needed to identify candidates for hard-to-fill positions.

Staffing is placing the right people in each position to provide the organizational capability needed to create a competitive advantage within the segments of the construction market deemed to be strategically important. There are three major components to the staffing function: planning, recruitment, and selection. *Planning* is forecasting the company's future human resources needs in the context of the strategic plan and the anticipated personnel vacancies, either from promotion, retirement, resignation, or layoff. *Recruitment* is attracting applicants for existing or future requirements. *Selection* is screening the applicants and selecting people for vacant positions.

Human resources planning involves assessing

- the firm's anticipated need for personnel,
- the skill sets qualified applicants must possess,
- the anticipated demand for personnel with those skills, and
- the forecast supply of qualified applicants.

The relevant personnel-recruiting market is defined by occupational skills and by geography. The demand from competitors must be considered when selecting the recruiting market, whether it be internal, local, regional, national, or international. The population of potential applicants is estimated in terms of both quantity and skill level.

Alternatives should also be considered while selecting; namely,

- full-time employees,
- temporary employees,
- part-time employees, or
- employees who share the jobs.

The last issue to be determined in developing the human resources plan is the designation of the selection authority for each position. The immediate supervisor typically performs this function, but someone higher in the organization may be designated as the selection authority.

One of the issues to be considered in the human resources planning process is whether or not to recruit internally. Internal recruiting is less costly than external recruiting because there is no need

- to advertise the positions outside the company,
- to send recruiters to recruiting activities, or
- to reimburse prospective employees for their travel to interviews.

Because internal applicants have performance records, the selection authorities will have a better knowledge of the internal applicants' skills and abilities than they will of those of the external applicants. Internal recruiting also enhances employees' morale because it provides career progression within the company. Internal recruiting, however, perpetuates existing business practices, because it eliminates the opportunity to bring in new people with fresh perspectives. The best practice is to conduct both internal and external recruiting and to select the people who best meet the job specifications.

External recruiting sources may be employee referrals, union referrals, advertising in newspapers or on company Web sites, employment agencies, search firms, and campus or vocational school visits. Most construction firms require each applicant to submit a standard application form, such as the example shown in Figure 7.3. In addition, resumes are usually required.

Once the applications have been received, the selecting authority must decide to whom to offer each position. The information used in making the decision is generally

- the information contained on the application form,
- any letters of recommendation provided,
- the applicant's resume,
- the contacts with listed references, and
- the applicant interviews.

The written documents are typically used to screen the candidates and select the most qualified individuals. The top ranked candidates are then interviewed to assess their qualities and qualifications for the position and to provide them information regarding the position.

Interviews must be planned carefully to ensure that each candidate is evaluated using the same criteria. The recommended approach is to develop an interview checklist similar to the one illustrated in Figure 7.4 and to use it for all interviews for the position to be filled. This will provide a tool to assist in comparing the interviews of the applicants under consideration. Interviews may be conducted by one person or by a panel. The objective of most interviews is to assess the applicant's

- job skills,
- ability to work with others, and
- motivation for the open position.

People skills and motivation are typically the most difficult attributes to assess. Once new employees are selected, they must be oriented regarding company policies and the specific jobs they are to perform as members of the company team.

A typical hiring process is shown in Figure 7.5.

FIGURE 7.3 Employment Application Form

Western States Construction Company
Employment Application Form

ALL INFORMATION FURNISHED IN THIS APPLICATION MUST BE ACCURATE AND COMPLETE (IN INK). ANY FALSE STATEMENTS OR OMISSIONS OF ANY KIND ARE GROUNDS FOR DENYING EMPLOYMENT.

IDENTIFICATION

Name: Last _____ First _____ Middle _____

Social Security Number: _____

Home Address: _____ _____ _____ _____
                Street   City   State   Zip

Home Telephone (Area Code and Number): _____

Work Telephone (Area Code and Number): _____

U.S. Citizen: Yes _____ No _____

If No, Type of Visa and Expiration Date: _____

Alien Registration Number: _____

EMPLOYMENT INFORMATION

Are you a former employee of Western States Construction Company? Yes _____ No _____

Former Western States Construction Company Employment: From _____ to _____
                                                                                        Month/Year   Month/Year

Employment Desired: Full Time _____ Temporary _____ Summer _____

Date Available: _____
                Month/Year

Expected Salary/Wage: _____

Will You Accept Overtime Work, if Offered? Yes _____ No _____

EDUCATION

Circle Highest Grade Completed: Elementary   Secondary   Advanced
                                                    1 2 3 4 5 6   7 8 9 10 11 12   13 14 15 16 17 18

Vocational Training/Special Courses of Study:
_____
_____
_____

List Any Professional Certificates or Licenses You Have Obtained:
_____
_____
_____

(continued)

List All Post High School Education Below. Show Degrees Received or Expected and Date.

School Name City and State	Dates Attended Month/Year	Major Minor	Degree Date (Month/Year)
	From		
	To		
	From		
	To		
	From		
	To		

EMPLOYMENT HISTORY

BEGINNING NOW, ACCOUNT FOR EVERY MONTH OF EMPLOYMENT OR UNEMPLOYMENT WITHIN THE LAST 48 MONTHS (4 YEARS). A RESUME MAY BE ATTACHED, BUT THIS SECTION MUST BE COMPLETED.

Firm Name: _____

Address: _____ City: _____ State: _____

Position Title/Duties:
_____
_____
_____

Date Employed: From _____ to _____
                       Month/Year   Month/Year

Reason for Leaving:
_____

Firm Name: _____

Address: _____ City: _____ State: _____

Position Title/Duties:
_____
_____
_____

Date Employed: From _____ to _____
                       Month/Year   Month/Year

Reason for Leaving:
_____

(continued)

FIGURE 7.3 (*continued*)

Firm Name: _____

Address: _____ City: _____ State: _____

Position Title/Duties:
_____
_____

Date Employed: From _____   to _____
                    Month/Year   Month/Year

Reason for Leaving:
_____

REFERENCES

PLEASE PROVIDE 3 REFERENCES WHO ARE KNOWLEDGEABLE OF YOUR QUALIFICATIONS.

Name: _____ Years Known: _____
Occupation: _____ Employed By: _____
Address: _____ City: _____ State: _____
Telephone Number (Area Code and Number): _____

Name: _____ Years Known: _____
Occupation: _____ Employed By: _____
Address: _____ City: _____ State: _____
Telephone Number (Area Code and Number): _____

Name: _____ Years Known: _____
Occupation: _____ Employed By: _____
Address: _____ City: _____ State: _____
Telephone Number (Area Code and Number): _____

SIGNATURE

THIS APPLICATION IS COMPLETE AND ACCURATE TO THE BEST OF MY KNOWLEDGE.

Signature: _____ Date: _____

(*continued*)

Interview Checklist

Name of Applicant: _____ Send Rejection Letter _____
Interview Location: _____ Second Interview _____
                                              Hold for Future Review _____

Date Available: _____
Location Preference: _____

Interview Rating:

Category	Unsatisfactory	Below Average	Average	Above Average	Outstanding
Physical—General impression, apparent energy, grooming, and speech					
Mental—Ability to organize thoughts, self-expression, and comprehension					
Maturity—Emotional stability and apparent leadership ability					
Motivation—Realism of aspirations and positive motivation goals					
Education—Good sense of direction and consistency					
Work Experience—Career growth and general quality					
Overall Rating					
	1	2	3	4	5

Comments:

Date: _____   **Overall Rating:**
Interviewer: _____

FIGURE 7.4 Sample Interview Checklist

EMPLOYMENT MANUAL

A construction company should create an employment manual that describes its employment policies and procedures. An example of an employment manual for a hypothetical company, Western States Construction Company, is shown in Appendix D. Employers can be bound by statements in employment manuals, if the manuals promise specific treatment in specific circumstances. Statements in employment manuals have been determined by courts to constitute promises. To avoid creating an employment contract, employment manuals should contain a disclaimer noting that the manual is

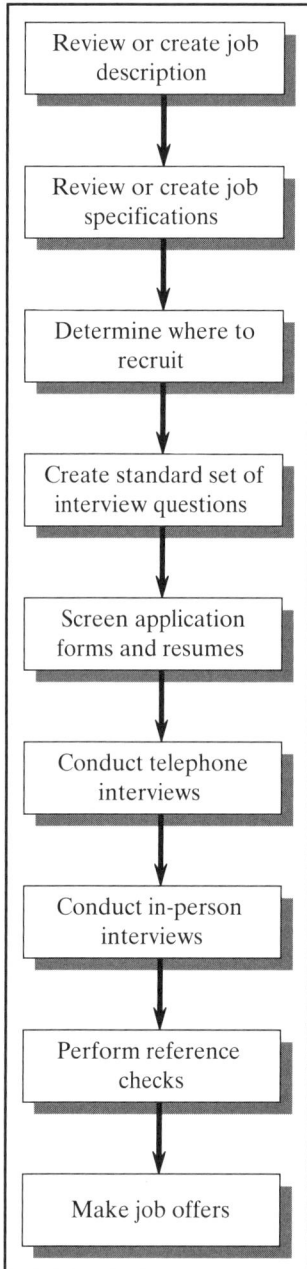

FIGURE 7.5 Typical Hiring Process

not intended to become a part of the employment relationship, but that the manual is instead a general statement of the employer's policy. The employer should reserve the right to alter the employment manual and change policies or promulgate new policies at its discretion without notice to the employees.

Employers should develop an employee acknowledgment form and require each employee to read the employment manual and submit a signed acknowledgment form.

CHAPTER 7　　Human Resources Management　　**149**

Employment Manual Acknowledgment Form

The employment manual describes important information about Western States Construction Company, and I understand that I should consult the Human Resources Department regarding any questions not answered in the manual. I have entered into my employment relationship with the Company voluntarily and acknowledge that there is no specified length of employment. Accordingly, either the Company or I can terminate the relationship at will, with or without cause, at any time.

　　Since the information, policies, and benefits described here are necessarily subject to change, I acknowledge that revisions to the manual may occur, except to the Company's policy of employment-at-will. All such changes will be communicated through official notices, and I understand that revised information may supersede, modify, or eliminate existing policies. The Company reserves the right at its discretion, to revise, amend, enforce, or even rescind the manual and/or individual policies at anytime. Furthermore, I acknowledge that this manual is neither an expressed or implied contract of employment nor a legal document.

　　This certifies that I have received a copy of the Employment Manual for Western States Construction Company. I acknowledge my responsibility to become familiar with and adhere to the Company's policies and work rules that are contained in this manual and those that have been explained to me at orientation sessions, meetings, or through postings. I agree to conform to the instructions, rules, and policies of the Company.

　　I understand that my employment is at will and can be terminated at any time, with or without cause and with or without notice, at the option of either the Company or myself. I understand that no representative of the Company has any authority to enter into any agreement for employment for any specified time, or to make any agreement contrary to the foregoing.

　　I understand that the policies in this manual supersede all previous policies, contracts, communications, and/or practices, written or verbal. I also acknowledge that I, as an employee of Western States Construction Company, have an affirmative obligation to notify my employer of any knowledge or information that I have relating to on-the-job safety violations committed by any employee at any time.

_____　　　　　　　_____
Employee Signature　　　　　　　　　　　　　　　　　　　Date

_____
Employee Name (Typed or Printed)

FIGURE 7.6　Sample Employee Acknowledgment Form

This should be done during the employee's initial orientation to the company, and whenever the manual is updated or modified. This will provide a record that all employees are aware of the contents of the employment manual. Some companies require the manual to be reviewed annually, while others require a review only when the manual is changed. Figure 7.6 is an example of an employee acknowledgment form.

EMPLOYEE DEVELOPMENT

Creating an environment where employees can develop their skills and have upward mobility opportunity is an essential element for employee retention and maintenance of competitive advantage. It is an important element to improve the worth of individual employees to the company. Individual employees may have different career aspirations and goals. As a part of a company mentoring program, individual development plans should be created for each employee. Investing in employee development is investing in the future of the company. Company leaders must ensure that individual managers within their companies understand their critical role in employee development and retention.

There are two important aspects to employee development. The first is skill-enhancement training designed to improve employee performance in his or her current position. The second aspect is development training to provide employees with the necessary skills to allow them to compete successfully for higher-level positions within the company. Both aspects are important. The first step in developing an employee development training program is to assess the needs of each employee for skill-enhancement training and for professional development. Next, specific training programs are either developed or identified. Employees are sent to specific training programs, and their performance is monitored to ascertain the effectiveness of the individual training programs. An individual development plan should be developed for each employee as a part of the performance evaluation process. This plan should be developed jointly by the employee and his or her supervisor. A sample format for an employee development plan is shown in Figure 7.7. Training needs should be prioritized, because most construction firms have limited training budgets.

Managers often have difficulty thinking of on-the-job development activities. To achieve the goal of continued employee growth and development, managers should help employees identify specific development objectives and formulate action plans to meet them. By doing this, managers help employees capitalize on their strengths while simultaneously addressing their development needs in a positive manner. Through helping employees to establish clear and specific goals, managers can tailor their developmental activities to employees' varied needs and learning styles. This also makes it possible to review and recognize change in an employee's performance that will lead to recognition, additional compensation, or promotion. Managers also gain the benefit of increasing their own effectiveness as employees are better able to handle additional or more complex tasks, and managers are freed up to focus on broader issues.

It is important that employees focus their efforts on priority areas or on changes that are most beneficial to both themselves and the company. A manager's role is first to help employees identify their own strengths and to select one to three of their most important strengths for greater use. To make the choice, managers should consider the extent to which the greater use of that strength will

- Benefit the company
- Increase employee satisfaction in the current job
- Prepare employees for their next positions
- Help employees develop their weaker areas

Employee development plans should be established and managed in much the same way as any business project. Specific objectives need to be established, and specific action plans need to be developed for each objective. Some issues that should be considered are

- What does the employee need to know to broaden the use of his or her skill?
- What complementary skills would develop the use of this skill?
- What can be done to help the employee use this skill differently?

Employee Development Plan

Employee Name: _____

Section One: (Employee Section)
(Employee provides an assessment of his or her current job performance, identifies educational needs for current position, identifies his or her career goals for the next five years, and identifies educational needs to prepare for future position.)

Section Two: (Supervisor Section)
(Supervisor provides an assessment of the demonstrated effectiveness of the employee.)

Section Three: (Goals Section)
(The employee and supervisor jointly identify training and development goals for the coming year.)

Employee Signature: _____ Date: _____
Supervisor Signature: _____ Date: _____

FIGURE 7.7 Sample Employee Development Plan Format

Target completion dates for each employees' action plans should be established to help employees make steady progress toward their objectives. Finally, supervisors should help their subordinates accomplish their development plans.

A second, and perhaps more obvious, way of improving employees' overall effectiveness is to work on their development needs. Development needs actually may be skills or behaviors they have to learn and do; others may be ones that they simply have to do. It is important for managers to focus development efforts on specific priority areas in which changes would be most beneficial. It is also important to note that employees will require different levels of attention and involvement depending on experience levels.

To keep employees on track, managers need to give frequent feedback. Feedback is essential for monitoring progress; it not only helps employees correct mistakes before they become serious problems, but also reinforces positive behaviors and encourages the development of desirable work habits. Feedback should be given in a timely manner. Start with positive feedback, and then, if warranted, discuss any areas in which improvement is needed. Supervisors should communicate a willingness to provide feedback. This will convey the desire to help and advance the team environment that is critical to employee satisfaction and company success.

Employees acquire skills gradually and begin using them in small, manageable steps. This is why managers need to help employees create development plans with a series of smaller steps that lead to each objective. Specific action steps are easier to track, and they allow employees to feel encouraged when they accomplish them. Sometimes, employees are unclear about how their work ties into the organization, or they may be asked to do things that seem meaningless or unimportant. Tying individual goals and tasks to the company's objectives can help employees view their assignments as important. Managers need to ensure that they and their teams plan and act consistently with the company strategic plan.

Continuous learning is essential if a company is to remain competitive. Employee training and development should be considered an investment rather than an expense. The challenge is to create an environment in which employees consider continuous learning as personal investment and not just time away from work. Companies that invest in the education of their employees and take advantage of the latest technologies realize increased business and profitability. A typical company budget for training and development is about 2 to 3 percent of the company payroll. Employees with updated skills and expanded knowledge enable a construction company to adapt to market changes quickly.

LEADERSHIP DEVELOPMENT

Construction companies need leaders at all levels, from the project superintendent to the company president. This is a strategic issue for company leaders. Construction company leaders today have to grow, develop, and renew their leadership talent in order to ensure the future success of their industry. The key is to identify individuals who are motivated to seek leadership positions, and then develop their leadership skills through education and experience. Good leaders are those who can inspire and ensure quality work from others.

What employees are looking for in their leaders are basic principles put in practice, such as honesty, integrity, ethics, and caring. A good leader must also be motivational, dedicated, fair, and a good listener. There are three basic steps to leadership development:

- *Experience.* Potential leaders should be assigned a variety of assignments throughout the company to ensure that they have specific experiences that are orchestrated to create specific learning outcomes.
- *Assessment.* Key lessons should be extracted from each set of experiences and mentors or coaches should provide feedback on how well the potential leaders performed.
- *Preparation.* The feedback received from mentors or coaches during the assessment becomes a springboard for preparation for the next experience.

A critical element in leadership development is to identify a mentor or coach for each prospective leader. It is vital to learn from the experience of others and apply that knowledge to future decisions. Most company leadership development programs are a blending of mentoring and participation in formal education programs. Another strategy is to encourage employees to participate in industry associations, such as the Associated General Contractors of America or the National Association of Home Builders. The benefit is realized as an individual moves through a variety of leadership positions in these organizations and establishes contacts outside the company.

Senior managers must make a commitment to identifying and developing a company's future leaders. The secret is to identify ambitious employees who aspire to leadership positions and craft leadership development programs to allow these people to acquire a variety of experiences. These future leaders will need to develop their ability to deal with rapidly changing business conditions and the expectations of younger employees. They also need to develop an ability to instill a sense of shared mission in their subordinates.

SUCCESSION PLANNING

Succession planning is identifying and developing future leaders, and perhaps owners, for a construction company. This is a critical and time-consuming responsibility of the current company leadership. For most construction companies, this is an eight- to ten-year process, with three years needed to identify the individuals, and five to seven years for leadership development. The sustainability of a company relies on the management developing its replacements. This requires a continuous process of leadership identification and development. A good succession plan enables a company to align the right employee with a position and develop the skills necessary to be successful when the position becomes vacant.

In small companies, succession planning is critical because the unanticipated departure or illness of a company leader can set the company foundering. Many owners of small companies are so consumed by the daily pressures of running a business that they fail to plan for the future. In doing succession planning, business owners need to determine whether they wish to

- keep the business in the family,
- recruit an outside manager to run the company,
- sell the company to key executives, or
- put the company on the market.

Typical succession planning involves several steps.

- First, anticipate management needs based on factors such as planned expansion or anticipated retirements.
- Next, review the company's inventory of management talent to select candidates for leadership positions.
- Then create replacement charts that summarize each candidate's strengths and development needs.

- Management development is then implemented, using such techniques as job rotation, coaching, and formal education.
- New leaders are then selected, and a transition plan is developed and implemented.

To develop future company leaders, it is important to identify leadership potential early and get them experienced in every division of the company. During the identification and development process, the current leaders need to analyze and concentrate on any weaknesses that the selected individuals may possess. Any weak areas need to be strengthened by education and experience. Prior to assuming their new responsibilities, the new leaders should have an opportunity to work with and shadow the individuals that they are to replace.

PERFORMANCE MANAGEMENT

Performance management involves the establishment of performance standards for each position and the creation of a performance appraisal system. Performance standards are developed jointly by the employee and his or her supervisor. An example of a set of performance standards for an administrative assistant is shown in Figure 7.8. A performance appraisal is a written evaluation of an employee's performance during a specific review period, typically one year. Forms similar to the one shown in Figure 7.9 are used to document this evaluation in an objective manner.

FIGURE 7.8 Sample Performance Standards Form

Performance Standards for Administrative Assistant

Name of Employee:
Supervisor:

Performance Standards:

1. Receive visitors, record their presence, and direct to appropriate office with no more than one error per month.
2. Receive telephone communications, electronic mail, and written documents and direct to appropriate individual with no more than 2 errors per month.
3. Maintain project files and misfile no more than 2 documents per month.
4. Prepare final copies of written communication with no more than one error per week.
5. Maintain time sheets for hourly employees without error and submit on time to payroll office.
6. Ensure reception area in project office is well organized and presents a professional image to visitors.
7. Keep office reproduction and facsimile machines operational by contacting appropriate vendor to perform any needed repairs.
8. Maintain a 30-day supply of consumable office supplies.

Employee Signature: _____ Date: _____
Supervisor Signature: _____ Date: _____

FIGURE 7.9 Sample Performance Appraisal Form

Western States Construction Company
Annual Performance Appraisal

Name of Employee: _____

Name of Supervisor: _____

Evaluation Period: From _____ to _____

For each applicable performance area, mark the box that most closely reflects the employee's performance using the following scale: 1 = unacceptable; 2 = needs improvement; 3 = satisfactory; 4 = above average; and 5 = outstanding

Performance Area	1	2	3	4	5
Ability to make job-related decisions					
Accepts responsibility					
Attendance					
Attitude					
Cooperation					
Dependability					
Effective under stress					
Initiative					
Leadership					
Operation and care of equipment					
Quality of work					
Safety practices					
Technical abilities					

Job Strengths: _____
_____

Areas for Improvement:
_____
_____
_____

Training Needed:
_____
_____
_____

Supervisor Comments:
_____
_____
_____

Employee Comments:
_____
_____
_____

Employee Signature: _____ Date: _____

Supervisor Signature: _____ Date: _____

The appraisal is prepared by the employee's supervisor and describes how well the employee has performed the tasks identified in the performance standards during the review period. The appraisal should focus only on performance attributes that relate to organizational success. Appraisals are often used to determine compensation increases, annual bonuses, and employee training needs. The appraisal system works best when there is frequent feedback and mentoring provided by the supervisor to the employee throughout the review period, not just when the appraisal form is completed. Training needs identified during the appraisal process should be included in the employee development plan discussed in the section "Employee Development."

COMPENSATION AND EMPLOYEE BENEFITS

Compensation levels are set for each position based on the relative worth of the position to the company and in the external labor market. Compensation levels must be set high enough to attract prospective qualified applicants and to provide incentives for experienced personnel to remain with the company. Compensation levels typically increase with experience and increased responsibility and at times when there is a shortage of qualified people in the labor market.

Annual compensation is typically composed of two components: base compensation and pay for performance. Base compensation may be set as an hourly wage rate or a monthly salary rate. Factors considered in establishing the base compensation are

- Skills necessary to perform the duties of the position,
- Effort required to perform the duties of the position, and
- Working conditions.

Salaried employees generally are exempt from the Fair Labor Standards Act and do not receive extra compensation for working overtime (beyond 40 hours per week). Nonexempt employees receive overtime compensation. Pay for performance may include individual incentive pay, group incentive pay, or profit sharing. Benefits typically are items such as health insurance, life insurance, worker's compensation, unemployment insurance, and retirement programs.

In terms of individual incentives (such as bonuses), if the requirements are too high or should expectations not be reinforced by managers, old patterns of performance will tend to resurface despite incentives. Managers who are able to motivate employees to change patterns of performance and remain with the company are more likely to increase job satisfaction and involvement than compensation alone. The expected relationship between the performance of a company's employees and the incentives the company is willing to provide and the perceived fairness of the incentives by employees are major factors in an incentive program.

EMPLOYEE RETENTION

The primary reasons why people leave a company are

- Lack of opportunity for advancement
- Dissatisfaction with management
- Dissatisfaction with compensation or benefits

- Conflict with coworker or manager
- Lack of recognition or appreciation
- Uncomfortable work environment

A way for managers to directly influence and build employees' commitment to a company is to involve them in as many decision-making processes as possible, providing ownership and accountability for their actions, and incentives for their success. Every decision involves an element of risk. Sometimes, however, sound decisions or decision alternatives are discarded because they appear too risky or because the decision maker feels uncomfortable with unproven alternatives. Calculated risk taking implies that a decision is made with a thorough understanding of the potential risks and benefits involved. The ability to recognize and take calculated risks is a skill required of all managers. By helping employees build these skills, managers provide employees with opportunities for growth that will increase the employees' ties to the company, thus keeping them focused on their roles within the company instead of searching for alternatives. The top motivators to retain valued team members are

- Importance of the work to be performed
- Appreciation of others
- Interesting work
- Enjoyable coworkers
- Expectation of financial reward

It is important for employees to have a vested interest in the success of the company. Many successful companies foster a sense of community by sponsoring fun events such as golf, baseball games, skiing, and picnics. Good internal communication is also critical. An employee newsletter is a good tool, either hard copy or electronic, or both.

If employees become dissatisfied, they will search for alternatives, compare these alternatives with their current situation, and then leave the company should one of the alternatives be perceived as more rewarding. The message for company leaders is that the way to keep employees satisfied is to understand what drives them, and through this understanding, implement strategies to increase their resistance to alternatives (employment) and respond to their needs. In doing so, the company should realize direct economic gains from the reduction of operating costs and from the avoidance of lost productivity. The lost productivity caused by turnover is evident in the inability of the company to handle the abandoned responsibilities in an efficient manner, the period of time in which a new employee comes up the learning curve, and the managerial oversight and participation that is required in the training of a new employee.

UNION RELATIONS

Union relations in the construction industry are quite different from those found in many firms in the manufacturing industry. Rather than negotiate a single contract with one union that represents all nonsupervisory personnel, union construction firms negotiate separate contracts with each craft union represented in their workforce, such as carpenters, laborers, operating engineers, and teamsters. The contracts define the working conditions for all members of the union who are employed by the construction company that signs the union contract. To conduct collective bargaining with each union, most unionized construction firms join trade organizations, such as Associated General Contractors of America, National Electrical Contractors Association, and Mechanical Contractors Association of America, who bargain on behalf of member firms.

Foremen generally retain union membership, but project managers usually do not. Superintendents may or may not retain union membership. It depends on their career path. Some superintendents started as craft labor and worked their way up. These individuals may choose to retain union membership. Others may have started as project or field engineers and never joined a union. Working conditions, including compensation, are prescribed in the individual union contracts

Construction firms are not allowed to form closed shops, in which only union members can be employed. In most states, they may form union shops, in which a new employee is required to join the union within a designated period or lose his or her job. In those states that have passed the right-to-work legislation, union shops are not allowed.

The major difference between union firms and nonunion firms is the training of new construction craftsmen. The unions have comprehensive apprenticeship programs for training entry-level personnel. Union construction firms needing additional skilled personnel submit their requests to the union hiring halls. Nonunion construction firms need to recruit to meet their personnel requirements and then develop skill-enhancement training programs to ensure that the new entrants possess the necessary technical skills before using them on a construction project.

SAFETY AND WELLNESS PROGRAMS

During the normal course of construction, employees are routinely exposed to various risks that can result in accidents. Unless measures are taken to preserve/protect employee well-being, accidents will occur. The provision of such protection requires a significant commitment of time, money, resources, and staff to manage, preserve, and protect employees in the most effective manner. Managers have a responsibility to keep employee work areas free from recognized hazards that might cause injury or harm to employees or property and to enforce the safety rules set for employees in the company's safety program. They also have the responsibility to make employees aware of the company's safety program's content, to communicate its goals, and to follow up with training and feedback to employee questions.

At a minimum, a company's safety program should

- Establish responsibility and accountability for proper implementation and enforcement of accident prevention policies.
- Provide policies and schedules for updating supervisors' skills and for allowing key employees to gain specialized safety and health training as appropriate.
- Be reviewed and updated annually to address changes in Occupational Safety and Health Administration (OSHA) standards.

Senior and executive managers should actively promote, facilitate, and enforce a company's safety program by

- Evaluating current safety policies and monitoring compliance with applicable federal, state, and local regulations.
- Giving the authority to lower level managers and frontline supervisors to administer and enforce safety programs—including the authority to stop work when unsafe conditions are encountered.
- Reviewing all accidents/incidents monthly.
- Reviewing corporate safety statistics at predetermined intervals.
- Actively seeking feedback on and implement improvements to safety training programs and methods to provide employees with a safe(r) work environment.

Project managers and field supervisors should

- Conduct field safety audits and inspections
- Perform prejob hazard analysis
- Support field superintendents in implementing safety programs
- Monitor safety training for field personnel
- Facilitate proactive field safety measures
- Review all accidents/incidents monthly
- Recommend corrective/preventative actions

Construction firms are required to provide worker's compensation insurance for their employees. This is a no-fault insurance that provides compensation to an employee who is injured on a job site. It covers medical costs, compensation if the employee is not able to return to work, and retaining for workers who are not physically able to perform their former jobs. Some states have a government monopoly, and the insurance must be purchased from a government agency. In other states, the insurance may be purchased from an insurance company. Some large construction firms have chosen to become self-insuring and have hired an administrator to administer their program. Not only is there a significant cost to having accidents on construction sites, but there is also a significant morale issue. People working on a job site want to feel good about working in a safe environment.

The purpose of the workers' compensation system is to provide a method to restore an injured worker as nearly as possible to the pre-injury earning capacity and potential. It is to everyone's advantage for an injured worker to return to work as soon as possible after injury, within medical restrictions, because returning to suitable work helps employees recover from injuries more readily, helps employers gain lost productivity, helps lower compensation costs, and there is less dependency on other types of assistance. It is helpful if employers have some type of limited duty, or alternate work, to help workers gradually get used to being back in the workforce.

Companies that have proactive return-to-work programs have

- A lower rate of lost workday cases
- A reduction in worker's compensation claims incidence
- Fewer lost workdays per 100 employees

A proactive program includes planned, coordinated, and supportive company-based interventions for assisting injured workers at the onset to return to work. Employees who participate in return-to-work programs return to work successfully, typically at their pre-injury pay rate. The programs reduced claims, lowered premium and litigation expenses, and improved the employer's public image.

While laws may, and most likely will, vary for each state, injured employees generally have the following rights:

- An injured employee may have the right to receive benefits.
- An injured employee may receive benefits regardless of who caused or helped cause the injury. An injured employee does not have a right to benefits if
    - the employee injured himself or herself intentionally,
    - the employee was injured while voluntarily participating in an off-duty activity, or
    - the injury occurred during horseplay or fighting initiated by the injured employee.
- An injured employee has the right to receive the medical care reasonable and necessary to treat a work-related injury or illness.

Worker's compensation only covers injuries sustained on the job site or associated with the work. Many construction firms have determined that wellness programs are effective in controlling the increased cost of employee medical insurance. These companies have determined that the programs can have a positive effect on employees by encouraging them to adopt more healthy lifestyles. A complete wellness program has three components:

- It helps employees identify potential health risks through screening and testing.
- It educates employees about such health risks as high blood pressure, smoking, poor diet, and stress.
- It encourages employees to change their lifestyles through exercise, good nutrition, and health monitoring.

Some companies pay for or subsidize employee membership in physical fitness centers or health clubs, while others construct such facilities at their company offices. Many companies cover the cost of smoking cessation and stress-reduction programs.

REGULATORY OVERVIEW

This section contains a brief overview of the primary Federal laws and regulations that affect company human resources management policies.

Affirmative Action

Companies must meet specific posting and reporting requirements to make certain that there is no discrimination in the employment or treatment of qualified employees based on any reason that is not reasonably related to job performance. A company's policy should require consistent and timely processing of all complaints. A good procedure has the following characteristics:

- Fairness and Objectivity
- Promptness
- Confidentiality

- Notice
- Thoroughness
- Finality

Age Discrimination

Discriminating on the basis of age in the workplace is illegal under the Federal Age Discrimination in Employment Act (ADEA). Under this law, there are some special limitations on who can sue. People under 40 years of age are not protected by the age discrimination in the workplace laws. If a company refuses to hire somebody because he or she is 39, and therefore "too young," that is not illegal. But if it because he or she is 40 and "too old," that is illegal. The ADEA's prohibitions against age discrimination apply to companies with 20 or more employees. The best shield a company can devise against liability in an ADEA case is to have neutral policies based on considerations of merit or seniority applied consistently.

Disabled Workers

With the passage of the Americans With Disabilities Act (ADA), job analysis has taken on an increasing importance. A job analysis is used to define the essential elements of the job, including the physical demands that the work requires. The ADA states that a company shall not discriminate against a qualified individual with a disability because of the disability in regard to job application procedures; the hiring, advancement, or discharge of employees; employee compensation; job training; and other terms, conditions, and privileges of employment. The Act defines a qualified individual with a disability as someone with a disability who, with or without reasonable accommodation, can perform the essential functions of the employment position that such individual holds or desires. A job analysis is usually required to determine what reasonable accommodations could be made for a disabled individual to perform the job.

The ADA cases involving HIV/AIDS continue to be litigated and companies encounter personnel decisions fraught with legal repercussions in managing employees with HIV/AIDS. Reassigning, accommodating, reducing or restructuring benefits, and dealing with other employees' fears and concerns raise legal and moral issues that challenge company leaders. Persons with HIV maintain protection under the ADA, and companies who do not address the issue effectively risk liability.

Discrimination

Direct or overt discrimination is any action that specifically excludes a person or a group of people from a benefit or an opportunity, or significantly reduces their chances of obtaining it, because a personal characteristic irrelevant to the situation is applied as a barrier. All companies should have written policies that prohibit discrimination on the basis of race, sex, color, religion, national origin, age, disability, and other categories.

These policies should

- Clearly state that discrimination won't be tolerated.
- Enumerate the types of discrimination that will not be tolerated.

- Describe the consequences for violating the policy.
- Be widely publicized in all appropriate employer materials, including job applications and announcements, employee handbooks, diversity materials, Web sites, intranets, and bulletin boards.
- Outline the procedure for workers faced with discriminatory practices.
- Provide for prompt and consistent investigation of any allegation of discrimination.
- Protect against retaliation.

Drugs in the Workplace

More job applicants are finding that part of the application process involves a preemployment drug test. Courts have consistently upheld the legality of requiring a preemployment drug test as a condition of employment. If a company plans to test current employees, the employer should have policies and procedures in place, including supervisory training and steps to take if there is a positive test. Postemployment testing can include random testing (for safety-sensitive positions), individualized suspicion testing, and postaccident testing.

Nonunion companies may require applicants and/or employees to take drug tests. In unionized workforces, the implementation of testing programs must be negotiated. The disciplinary consequences of testing positive need to be determined and are subject to collective bargaining.

Equal Employment Opportunity

Equal employment opportunity (EEO) means equal access to jobs and benefits and services for all employees and prospective employees in the workplace. EEO aims to ensure fair and equitable outcomes in all areas of employment that relate to recruitment, selection, access to information, supervision, and management. EEO is a policy that promotes selection by merit. EEO means that all people have the right to be considered for any job for which they are skilled and qualified, and that they will be judged for the job on the basis of merit. EEO laws were designed to promote

- Fair practices in the workplace.
- Management decisions being made without bias.
- Recognition and respect for the social and cultural backgrounds of all employees and customers.
- Employment practices which produce staff satisfaction, commitment to the job, and the delivery of quality services to customers.
- Improved productivity by guaranteeing that the best person is recruited and/or promoted, skilled employees are retained, and the workplace is efficient and free of harassment and discrimination.

Managers need to ensure that

- The work environment is free from all forms of harassment.
- Employees are provided with information that will assist them to carry out their duties.
- All staff members have an equal opportunity to increase their skills to meet work demands, to attend training courses, and to apply for all available jobs.

The EEO policy statements and practice procedures in employee handbooks satisfy both this objective and that of continuously reinforcing the company's commitment to EEO. The inclusion of an EEO complaint procedure in the EEO policy statement apprises employees of the company resources available to them for EEO-related complaints that may arise.

Preventing Violence in the Workplace

Construction companies are starting to recognize the enormity of the financial consequences associated with an incident involving workplace violence. The three most affected areas are costly litigations, lost productivity, and damage control. Lost productivity following an incident is frequently underestimated. Losses in productivity occur throughout the workplace for a period of time after the incident. Losses are caused by the unavailability of the killed or injured worker, the work interruptions caused by investigations and damage to the facility, the time lost by surviving employees talking about the incident, the decreased efficiency and productivity due to posttraumatic stress syndrome, and the time spent by employees in counseling sessions.

Some suggested techniques for minimizing the potential for violence in the workplace are

- Take action immediately, if an employee exhibits bizarre behavior.
- Before terminating a potentially violent employee, seek professional help.
- Develop a crisis action plan to address violence in the workplace.
- Make it known that threats of violence will not be tolerated.
- Beware of workplace romances.
- Limit public access to work areas.

Privacy Rights

Employee rights fall into three categories: the right to job security, the right to fair treatment by the employer, and the right to fair treatment in the workplace. Treating employees with respect and fairness is critical for two reasons. First, it establishes a company's reputation for fairness and impartiality. This reputation is carefully scrutinized by individuals both within and outside the company and is vital for attracting and retaining employees. The second and equally important reason is that identifying and safeguarding employee rights reduces the possibility of the company becoming embroiled in charges of discrimination, lengthy litigation, and costly settlements.

Racial and Ethnic Discrimination

Title VII of the Civil Rights Act of 1964 prohibits workplace discrimination based on religion, national origin, race, color, or gender. The law's prohibitions include harassment or any other employment action based on any of the following:

- Affiliation: Harassing or otherwise discriminating because an individual is affiliated with a particular religious or ethnic group.
- Physical or cultural traits and clothing: Harassing or otherwise discriminating because of physical, cultural, or linguistic characteristics, such as accent or dress associated with a particular religion, ethnicity, or country of origin.

- Perception: Harassing or otherwise discriminating because of the perception or belief that a person is a member of a particular racial, national origin, or religious group whether or not that perception is correct.
- Association: Harassing or otherwise discriminating because of an individual's association with a person or organization of a particular religion or ethnicity.

Employers must provide a workplace that is free of harassment based on national origin, ethnicity, or religion. They may be liable not only for harassment by managers, but also by coworkers or by nonemployees under their control.

Record Keeping

Personnel documents are those records that employers use, have used, or intend to use to determine an employee's qualifications for employment, promotion, transfer, additional compensation, disciplinary action, or discharge. However, employers may not gather or keep records of employees' nonemployment activities, such as an employee's associations, political activities, or publications unless

- The activities occur on the employer's premises or during the employee's working hours and interfere with the performance of job duties by the employees.
- The activities constitute criminal conduct.
- The activities may reasonably be expected to harm the employer's property, operations, or business.
- The activities could cause the employer financial liability.

Companies who do maintain personnel records regarding their employees must permit employees to inspect their personnel documents at least twice each calendar year at reasonable intervals. Companies must allow the inspections within seven working days after an employee makes a request unless, under the circumstances (for example, due to the business's workload), the employer cannot comply with the seven-day deadline. In that case, the company has an additional seven days in which to allow the inspection.

Religious Protection

Title VII of the same Civil Rights Act requires an employer to reasonably accommodate the religious practices of an employee or prospective employee, unless doing so would create an undue hardship for the employer. Some reasonable religious accommodations that employers may be required to provide workers include leave for religious observances, time and/or place to pray, and ability to wear religious garb.

As a general rule, companies may not regulate employees' personal religious expression on the basis of its content or viewpoint. Companies may, at their discretion, reasonably regulate the time, place, and manner of all employee speech, provided such regulations do not discriminate on the basis of content or viewpoint. In informal settings, such as cafeterias and hallways, employees are entitled to discuss their religious views with one another.

Because managers have the power to hire, fire, or promote, employees may reasonably perceive their managers' religious expression as coercive, even if it was not intended as such. Therefore, managers should be careful to ensure that their statements and actions are such that employees do not perceive any coercion of religious or nonreligious behavior (or respond as if such coercion is occurring), and should, where necessary, take appropriate steps to dispel such misperceptions.

Federal law requires a company to reasonably accommodate an employee's religious observances, practices, and beliefs. However, a company need not reasonably accommodate if the company can show that accommodation would cause an undue hardship on business.

Sexual Harassment

Sexual harassment has been defined by the United States Equal Employment Opportunity Commission (EEOC) to be as follows:

> Unwelcome sexual advances, requests for sexual favors, and other verbal or physical conduct of a sexual nature constitutes sexual harassment when submission to or rejection of this conduct explicitly or implicitly affects an individual's employment, unreasonably interferes with an individual's work performance or creates an intimidating, hostile or offensive work environment.

This definition has been further refined by numerous court cases to indicate that sexual harassment can occur in a variety of circumstances, including but not limited to the following:

- The victim as well as the harasser may be a woman or a man. The victim does not have to be of the opposite sex.
- The harasser can be the victim's supervisor, an agent of the employer, a supervisor in another area, a coworker, or a nonemployee.
- The victim does not have to be the person harassed but could be anyone affected by the offensive conduct.
- Unlawful sexual harassment may occur without economic injury to or discharge of the victim.
- The harasser's conduct must be unwelcome.

There are two legally recognized types of sexual harassment:

- Quid pro quo sexual harassment
- Hostile environment sexual harassment

Quid pro quo sexual harassment occurs when an individual's submission to or rejection of sexual advances or conduct of a sexual nature is used as the basis for employment decisions affecting the individual, or when the individual's submission to such conduct is made a term or condition of employment. Hostile environment sexual harassment occurs when unwelcome sexual conduct unreasonably interferes with an individual's job performance or creates a hostile, intimidating, or offensive work environment even though the harassment may not result in tangible or economic job consequences; that is, the person may not lose pay or a promotion.

An employer can be held liable for the creation of a hostile environment by a supervisor, by nonsupervisory personnel, or by the acts of the employer's customers or independent contractors if the employer has knowledge of such harassment and fails to correct it.

Summary

A construction company's human resources represent the social capital of the firm. Without an adequate organization composed of skilled team members, the company will not be able to develop the competitive advantage needed to be successful in the construction market. Two major challenges facing company leaders are how to attract, develop, and retain a skilled, motivated workforce and how to comply with the myriad of legal requirements. The company culture plays a critical role in employee retention.

Organizational design is the selection of the organization chart for the company and the determination of the number of people needed in each organizational element. Once the organization chart has been developed, the individual jobs are analyzed to determine a job description and a set of job specifications for each position. Staffing involves planning human resources requirements and selecting strategies for recruiting and selecting people. Application forms and interviews are generally a part of the recruiting and selection process. To provide a single reference of company policies, all construction firms should develop employment manuals.

Employee development programs are used for short term—skill enhancement training and for long-term professional development. Individual development plans should be developed for each employee to guide the annual investment in improving employee development. Leadership development programs need to be implemented to develop future leaders for the company. A succession plan needs to be developed to provide for the continuity of operations in the event that one or more company leaders depart.

Performance management involves establishment of performance standards for each position and creation of an appraisal system. Appraisals are often used to determine compensation levels, annual bonuses, and employee training needs. A major factor in employee performance is timely feedback from supervisors. Compensation levels are set based on the value of the position to the successful accomplishment of the company's strategic plan and the value within the external labor market.

Employee retention is greatly influenced by how employees are treated within the company. Supervisors play a significant role in keeping employees content to remain as members of the company team. The human resources management process for craftspeople in a union construction firm is governed by the terms and conditions stipulated in the individual union contracts. Safety and wellness programs are used to motivate employees to adopt healthy lifestyles and work safely on project sites.

To comply with Federal laws and regulations, company leaders should become familiar with the laws regarding affirmative action, disabled workers, discrimination, drugs in the workplace, equal employment opportunity, preventing violence in the workplace, and sexual harassment.

Review Questions

1. What is the relationship between a construction firm's human resources management plan and its overall strategic plan?
2. How does a company's decision-making process influence employee motivation?
3. What is meant by the term "company culture?"
4. How do company leaders influence the company culture?

5. What are the major tasks involved in organizational design?
6. What type of information is shown in the job description for a project superintendent?
7. What type of information is shown in the job specifications for a project superintendent?
8. What are the major tasks to be performed in the staffing function?
9. Why are a job description and job specifications needed to conduct external recruiting?
10. What are the advantages of internal recruiting?
11. What are the advantages of external recruiting?
12. What is the difference between the screening process and the selection process when selecting employees for vacant positions?
13. What is the employment manual, and why is it an important document?
14. What are the two major considerations in creating an employee development program?
15. What are individual development plans, and who prepares them?
16. What are the three basic steps in leadership development?
17. Why is succession planning a critical issue for leaders of construction companies?
18. What are the performance standards used for?
19. What is a performance appraisal used for?
20. What is the difference between compensation and employee benefits?
21. How do company managers influence employee retention?
22. How are craft unions organized in the construction industry?
23. Why are safety and wellness programs an essential component of human resources management?
24. What is the age threshold for protection under the ADEA?
25. Are companies required to hire disabled workers who apply for vacant positions?

Exercises

1. Prepare a job description and a set of job specifications for a project manager.
2. Prepare a set of performance standards for a project manager.
3. You have been asked to participate in interviewing new applicants for a vacant project manager position. What questions would you ask at the interview?
4. The senior officers of Northwest Constructors (Appendix C) are nearing retirement. What type of a succession plan do you suggest the company adopt?

Sources of Additional Information

Bernardin, H. John. *Human Resources Management: An Experiential Approach*, 4th ed., New York: McGraw-Hill, Inc., 2007.

Connor, Mary P. and Julia B. Pokoria. *Coaching and Mentoring at Work: Developing Effective Practice*, New York: McGraw-Hill, Inc., 2007.

Dressler, Gary. *Human Resources Management*, 11th ed., Upper Saddle River, N.J.: Prentice-Hall, 2008.

Gomez-Mejia, Louis R., David B. Balkin, and Robert L. Cardy. *Managing Human Resources*, 5th ed., Upper Saddle River, N.J.: Prentice-Hall, 2007.

Moran, John. *Employment Law: New Challenges in the Business Environment*, 2nd ed., Upper Saddle River, N.J.: Prentice-Hall, 2002.

CHAPTER 8

Information Management

INTRODUCTION

Just as a construction company needs financial resources to finance its business activity and human resources to provide customer service, it also needs information to support its decision-making processes and to provide historical records. The needed information must be accurate, accessible, relevant, timely, and complete. Thus, the collection, the analysis, the storage, and the retrieval of information are essential functions of a construction company. To perform this essential function, some construction firms are hiring chief information officers, whose responsibilities involve integrating information technology to meet company business objectives and to serve users within the company. Other construction firms have chosen to outsource this function to consulting firms. Which approach to select depends on the business strategy of the company.

If we look at a construction firm as a system, we see that the information infrastructure is a critical component, as illustrated in Figure 8.1. The company's employees are both collectors as well as users of information. They also may make adjustments to the company business processes and organizational structure. The business practices determine information requirements, as well as information collection techniques. The organizational structure influences the decision-making processes by defining relationships among the employees and establishing information-reporting procedures. To be effective, the company's business practices must be compatible with the company organizational structure and corporate culture.

An information management system collects, processes, stores, analyzes, and disseminates information as illustrated in Figure 8.2. The company's business practices define the information to be collected, and the company's employees exercise control by adjusting the processing mechanisms based on analysis of the output received. The information collected becomes the historical record that can be used to assess the accomplishment of the company's business plan and for making adjustments in resource allocation among the company's various departments.

Historically, much of the needed information was collected in paper form and stored in file cabinets for future reference and use. Today, most of the information is collected and stored in electronic form. The collection, the analysis, and the storage of information involve the expenditure of financial resources to hire staff to perform the tasks and to purchase equipment and software for their use. Therefore, company leaders must ensure that there is a business need for all the information collected and that the stored information is organized in a structure that facilitates efficient retrieval. The test of a good system is the ability of employees to locate and retrieve needed information quickly.

CHAPTER 8   Information Management   169

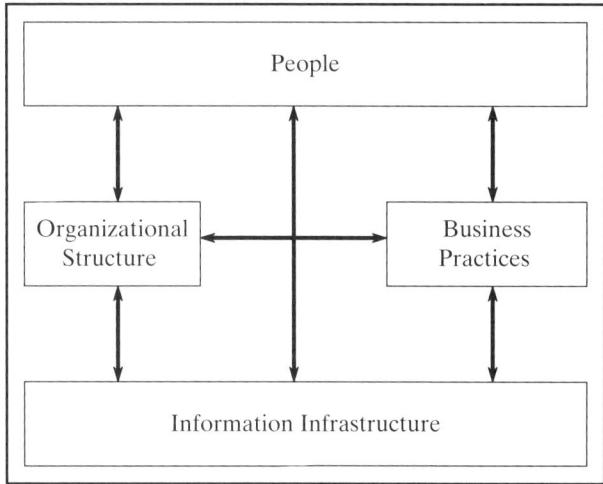

FIGURE 8.1 System View of a Construction Company

The major components of a modern company information management system are illustrated in Figure 8.3. Raw data are collected by company employees, and the data are organized and stored for retrieval within the information management system that is composed of hardware, software, and telecommunications networks. The **hardware** is the equipment used to operate the system, the **software** provides the desired functionality to the equipment, and the telecommunications networks provide the connectivity among the hardware components. Investments in information management technology should be addressed in the development of a company strategic plan. Such investments should be considered as ongoing regular expenses, rather than one-time costs. In addition, there will be a continual need to invest in employee information technology literacy training.

In this chapter, we will discuss the planning and implementation of an information management system. The capabilities desired in such systems typically include

- Ability to perform high-speed numerical computation
- Ability to provide fast, accurate, and inexpensive communication

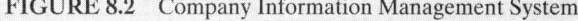

FIGURE 8.2 Company Information Management System

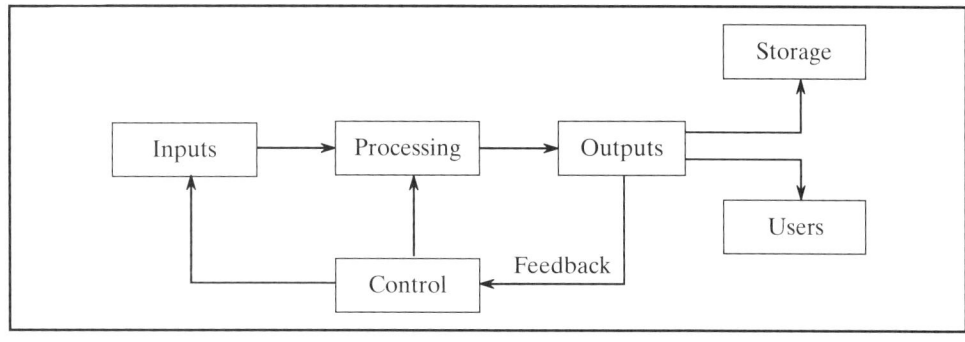

CHAPTER 8 Information Management

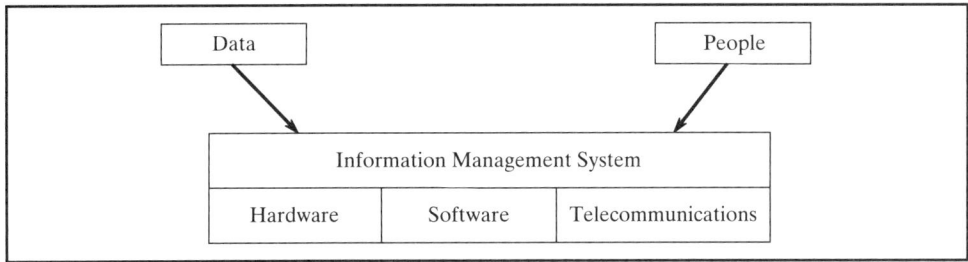

FIGURE 8.3 Components of Company Information Management System

- Ability to store information in easy-to-access format
- Ability to be accessed from multiple locations

In addition to implementing an information management system, company leaders must ensure that company employees are trained on its use and understand the importance of inputting accurate information.

INFORMATION REQUIREMENTS

The collection and management of information involves the expenditure of resources and should be undertaken only if the information is needed to support the company's business processes. The expanding capabilities of information technology have led to adjustments in the business processes of many companies. Information that does not support business processes should not be collected and maintained just because technology is available for its collection. The first step in information management planning is to map the company's business practices and identify the information needed for each process as well as the information needed for historical records. The information collected should be accurate, relevant, timely, and complete.

In general, the primary business processes of a construction company can be grouped into the following categories:

- Accounting processes
- Business development processes
- Contract management processes
- Cost-estimating process
- Customer relationship management process
- Equipment management processes
- Financial management processes
- Human resources management processes
- Procurement processes
- Project cost control processes
- Project management processes

The amount of detail required by company decision makers varies depending on their location within the company hierarchy, as shown in Figure 8.4. At the operational

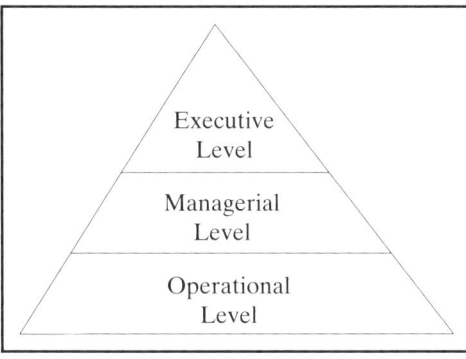

FIGURE 8.4 Decision-Making Levels within a Company

level where routine decisions are made and interactions with customers and suppliers occur, detailed information is needed. The functional managers at the managerial level need summarized information to execute their oversight responsibilities. Company executives need aggregated information focusing on the company strategic goals and objectives.

HARDWARE AND INFRASTRUCTURE

As shown in Figure 8.3, two of the primary components of a modern information management system are the hardware and the telecommunications networks. The primary hardware components are

- Input devices such as keyboards and scanners
- Output devices such as monitors and printers
- Storage devices such as hard drives, DVD, CD, and flash memory
- A central processing unit and random-access memory
- Connecting devices such as networks or routers

The backbone of a company's information management system is the computers and other electronic devices that are linked by telecommunications networks. The connections allow users to access information stored in multiple locations and to communicate with others. One type of such networks is known as a local area network, which allows computer users in a single area or department to share information and peripheral devices, such as a printer or a scanner. Application software may be stored on the local area network for use by multiple users. Such networks typically use computers, called servers, to deliver information and software to other computers linked in the network.

Company computers may also be connected to a global network environment, known as the Internet. Some companies have also developed **intranets**, which are web-based networks accessible only to company employees. In addition, some companies link portions of their intranets to those of their business partners over networks known as **extranets**. Companies often create project extranets to facilitate collaboration among project participants: owner, designer, subcontractors, and suppliers. Access to company intranets and extranets is usually controlled by the use of passwords, which will be discussed in the section "Security Systems." Web sites that provide access to company intranets and extranets are known as company portals.

It is recommended that purchasing and maintenance of hardware be based on overall company policies, rather than to let each department establish its own policies. This is to ensure that all hardware purchased is compatible with the company's information technology policies, is supported by appropriate maintenance contracts, and meets company information management requirements.

The telecommunications networks allow employees to telecommute to their office by allowing them to work from home or remote locations. Such networks allow access to company intranets and communication by electronic mail. Some construction companies purchase specialized equipment to create a capability for video-teleconferencing among people located at diverse locations. Such capabilities are used to hold virtual meetings, avoiding the need for employees to travel to a central location to attend a meeting.

SOFTWARE

Software allows the information management system to perform the many desired functions. Various types of software are shown in Figure 8.5. Application software is designed to support specific tasks or business processes, such as cost estimating or payroll. Operating system software controls the application software and manages how the hardware devices work. An example of an operating system software is Microsoft Windows. Utility software provides extra functionality to the operating system software. An example of a utility software is an antivirus software. A key business decision related to software is when to upgrade to newer versions. Software manufacturers continuously upgrade their products to remain competitive, but it is not necessary to purchase all

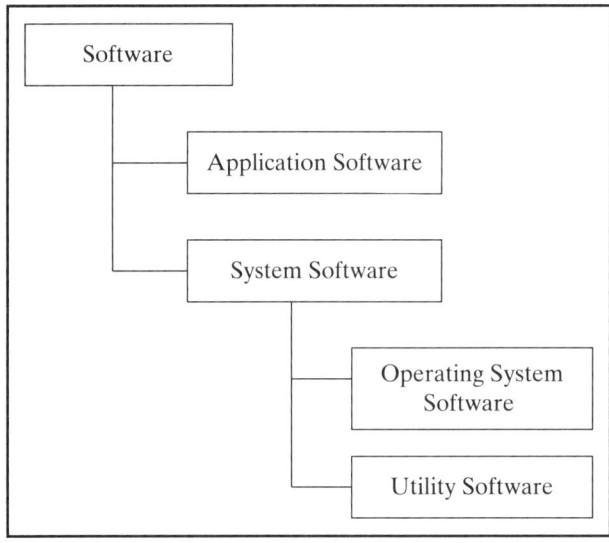

FIGURE 8.5 Information System Software

upgrades. Let the newer versions be tested by others and purchase them only if they offer significant performance enhancements.

A key part of the information management system is one or more databases, which are collections of related data that have been formatted in a manner to facilitate searches. A database management system is a software application (such as Microsoft Access) that is used to organize, store, and retrieve data from one or more databases. Report formats are created to compile data from databases and organize the data into useful formats that can be viewed and printed. Databases have become the electronic equivalent of filing cabinets with records and documents. Since databases are used to store essential company information, they must be properly designed and managed to support company business processes.

SECURITY SYSTEMS

Security systems are used to keep computers and information safe from unauthorized access and damage. Passwords should be required to use company computers and access company networks, such as local area networks or company intranets. Firewalls are usually created to limit access to company intranets or extranets. **Firewalls** are hardware devices with special software, which are placed between the Internet and company networks, as illustrated in Figure 8.6. Passwords are used to pass through the firewall to access the company intranet.

Company data should be backed up regularly and stored in multiple locations to protect its integrity. Data backup is like an insurance policy to allow restoration of company files in the event of an accidental corruption or a disaster, such as a fire in an office. Antivirus software should be installed to prevent virus attacks. Viruses may be received unexpectedly via electronic mail or Internet transactions and may corrupt the system software. To protect against such threats, antivirus software should be installed on all computers to detect the presence of viruses.

FIGURE 8.6 Company Firewalls

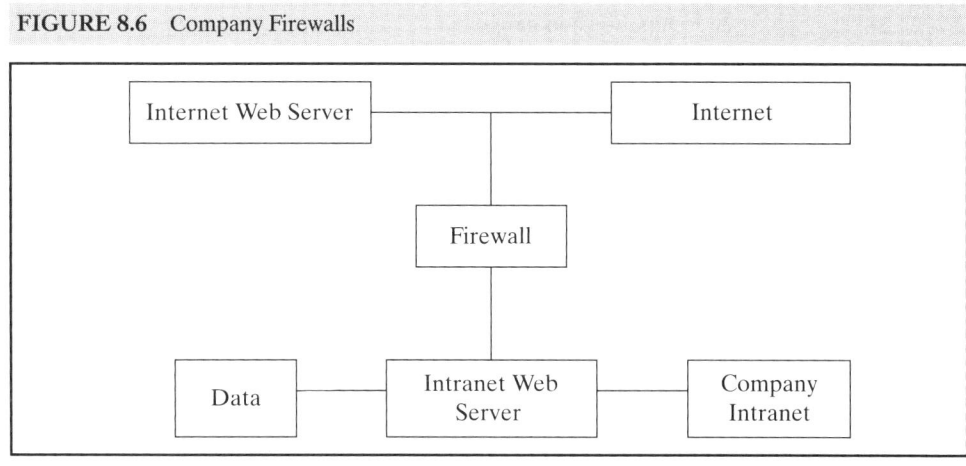

COMPANY SECURITY POLICY

All companies should have a published security policy to guide employees in the protection of company information resources and equipment. An example of such a policy is shown below.

Company Security Policy

Company Computers

- Company computers are to be used for business purposes only.
- Loading of any programs or downloading of any data onto company computers without company permission is prohibited.
- All data stored on company computers or servers is the property of Western States Construction Company.
- The transfer of any data from company servers or computers without a business purpose is prohibited.

Electronic Mail

- Our electronic mail system is for official business.
- We expect employees to honor our password protection system and not read other employees' e-mail. However, please remember that e-mail is neither private nor confidential. Any message that you send can be forwarded. Any e-mail including those deleted can be retrieved. All messages are company records and are the property of Western States Construction Company. We reserve the right to read, use, and disclose e-mail messages. For these reasons, e-mail should not be used for personal or private communication.
- When using e-mail, keep in mind that you are using company property. As a result, your comments must be appropriate to our business setting.

Internet

- Access to the Internet is provided for company business.
- Please keep in mind that Internet traffic can be monitored. Avoid Web sites that may contain information that would not be appropriate for our business. Be sure to avoid sites that include content that would be inconsistent with our policies prohibiting discrimination and harassment.

Misuse

- Improper use of company communication services or equipment is subject to discipline, including discharge. If you suspect any misuse of company systems, notify your supervisor.
- The company reserves the right to limit and/or terminate access to the electronic communications system at any time.

COMPANY WEB SITES

A company Web site is an important asset for establishing a presence on the World Wide Web. Web sites provide company information to interested people 24 hours per day, seven days per week. Having a company Web site is essential to creating a modern image that the company is technologically advanced. Prospective customers may use the Web site to examine the types of services offered and the types of projects the company undertakes. A good company Web site can be an effective recruiting tool for the human resources management staff. Company leaders, business development managers, and human resources managers should have contact information available on the company Web site. A well-designed Web site can significantly reduce the staff time spent on answering routine questions about the company. To be effective, the Web site needs to be maintained to ensure that all information contained is current and correct.

Summary

A construction company needs information to support its decision-making processes and to provide historical records. Thus the collection, the analysis, the storage, and the retrieval of information are essential business functions. A company information management system collects, processes, stores, analyzes, and disseminates information based on the company's business practices. The collection and management of information requires resources and should be undertaken only if the information is needed to support the company's business processes.

The major components of an information management system are hardware, software, and telecommunications networks. The primary hardware components are computers and other electronic devices that are linked by telecommunications networks. The computers may be linked to local area networks that allow sharing of information within a department, company Web-based networks known as intranets and extranets, and a global network known as the Internet. The software provides the necessary functionality to the hardware.

Security systems are used to keep computers and information safe from unauthorized access and damage. Passwords, firewalls, and antivirus software are the primary security tools. Company Web sites are used for establishing a presence on the World Wide Web. These sites provide essential information about the company.

Review Questions

1. What is the relationship between the design of a company information management system and the company's business practices?
2. How does a company's organizational structure affect the design of the company's information management system?
3. What are the major components of a modern information management system?
4. What are the three capabilities desired in most company information management systems?
5. What process would you use to determine what information should be collected by a company's information management system?
6. What are the four types of hardware components that may be used in a company information management system?
7. What are local area networks?
8. What is the difference between a company intranet and the Internet?
9. What is the difference between a company intranet and a company extranet?

10. What is the difference between application software and system software?
11. What are firewalls, and why are they used in company information management systems?
12. What are company Web sites, and why are they used?

Exercises

1. You are the chief information officer for a construction company that has a home office and four field offices. Each field office manages three to four separate projects. What combination of intranets, extranets, and local area networks would you establish to provide an efficient information management system for the company? Who would you allow to have access to the company extranets?
2. You have been asked to develop a database to support your company's contract management processes. What information do you suggest be collected in the database?
3. You have hired a consultant to develop a company Web site. What information do you want the Web site to contain? Why did you select this information?

Sources of Additional Information

Haag, Stephen, Maeve Cummings, and Donald J. McCubbrey. *Management Information Systems for the Information Age*, 5th ed., New York: McGraw-Hill/Irwin, 2005.

Jessup, Leonard M., and Joseph S. Valacich. *Information Systems Today*, Upper Saddle River, N.J.: Prentice-Hall, 2003.

Martin, E. Wainwright, Carol V. Brown, Daniel W. DeHayes, Jeffrey A. Hoffer, and William C. Perkins. *Managing Information Technology*, 4th ed., Upper Saddle River, N.J.: Prentice-Hall, 2002.

Mechanical Contractors Association of America Technical Committee. *Management Strategies for Information Technology*, Rockville, Md: Mechanical Contractors Association of America, 2003.

Turban, Efraim, Dorthy Leidner, Ephraim McLean, and James Wetherbe. *Information Technology for Management: Transforming Organizations in the Digital Economy*, 5th ed., Hoboken, N.J.: John Wiley & Sons, Inc., 2006.

CHAPTER 9

Total Quality Management

INTRODUCTION

Total quality management (TQM) is a management philosophy that focuses on continuous process improvement and customer satisfaction. It is a proactive management approach that seeks to prevent mistakes, rather than detect and correct mistakes. It uses some of the strategic management principles discussed in Chapter 5 and business development concepts discussed in Chapter 6. The objective is to determine customer expectations and to create business processes that meet or exceed those expectations every time.

Total quality management was introduced in the early 1950s in Japan by an American statistician, W. Edwards Deming. He believed that quality programs would not work unless the corporate culture was receptive and management practices and procedures were focused on quality rather than cost control. Introduced in the United States in the 1980s, TQM has been adopted by many manufacturing and service companies. Deming advocated 14 principles[1] to realign management practices for improving the quality culture of the organization (Figure 9.1), while warning of the following "seven deadly diseases" that obstruct the search for quality:

1. *Lack of consistency of purpose.* Leaders must create a culture that ensures continued emphasis on quality. It cannot be treated as just another program. Employees must believe that the quality culture is permanent.
2. *Emphasis on short-term profits.* Quality is an approach that is developed over time and is not focused on short-term financial performance.
3. *Evaluation of performance, merit rating, or annual performance review.* Evaluations that emphasize individual performance undermine the teamwork that is needed in quality-based organizations.
4. *Mobility of management.* If a manager perceives that the successful career track is a series of short stints in multiple positions, the focus will be on short-term results. A long-term commitment and perspective is needed to make quality work.
5. *Management by use of visible figures.* Customer satisfaction is a multifaceted concept that cannot be totally quantified.
6. *Excessive medical costs.* Operating costs are affected by the laws and social expectations of the countries in which the company is doing business. If a company has poor wellness and safety programs, it often incurs excessive medical costs and experiences high employee use of sick leave.
7. *Excessive liability.* There are legal costs that result from not producing quality products and services.

---

[1]W. Edwards Deming, *Out of Crisis*, Cambridge, Mass.: MIT Center for Advanced Engineering Study, 1986.

CHAPTER 9   Total Quality Management

---

1. Create consistency of purpose with a plan
2. Adopt the new philosophies of quality
3. Cease dependence on mass inspection
4. End the practice of choosing suppliers based on price
5. Find problems and work continuously on the system
6. Use modern methods of training on the job
7. Change from production numbers to quality
8. Drive out fear
9. Break down barriers between departments
10. Stop asking for productivity improvements without providing methods
11. Eliminate work standards that prescribe numerical quotas
12. Remove barriers to pride of workmanship
13. Institute vigorous education and training
14. Create a structure in top management that will push on the above 13 points every day

---

FIGURE 9.1 Deming's Fourteen Principles

Deming believed that improving quality would lead to a chain reaction within the company. Improved quality would lead to lower cost because of less rework, fewer mistakes and delays, and less material waste. Productivity would then improve because there would be less defective output, allowing a company to have greater market penetration because of lower cost and higher product quality. Greater demand for the company's products would result in higher profit margins.

Deming also created the plan-do-check-act cycle shown in Figure 9.2 to illustrate the continuous process improvement aspect of TQM. With the customer as the focus, the company plans for quality, implements those plans using a test sample (the do stage), reviews the results of the test (the check stage), and makes any necessary process changes before adopting the revised procedures as standard (the act stage).

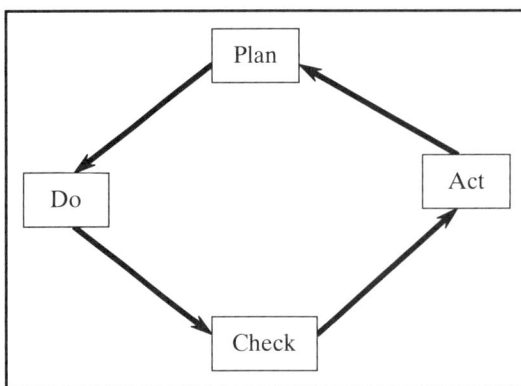

FIGURE 9.2 Deming Quality Improvement Cycle

The process is then reevaluated to identify additional potential improvements, and the cycle is repeated. Let's consider a construction example. United Constructors has determined that its process for managing shop drawings needs improvement. A team of five people is formed to review the process and make recommendations for improving it. The team conducts a detailed assessment of the current procedures and the expectations of all who participate in the process. Based on this assessment, process improvements are identified and implemented. The results of the process enhancements are measured, and the process is reevaluated to identify additional improvements.

Initially, TQM was introduced in the manufacturing industry, but the concept has now been widely adopted in the service industry also. The construction industry has been slow to adopt TQM, because of the short duration of most construction projects and the multiple organizations involved in performing the construction. However, TQM can be used to improve business processes both within a construction company and among the organizations involved on a typical construction project.

QUALITY CONCEPTS

Quality can be defined as meeting customers' requirements, needs, and expectations, the first and every time. The customer is defined as the recipient of the service provided by or the product produced by a management process, such as processing field questions or contract change orders. Customers may be internal or external to the construction firm. Internal customers are company employees who are users of company processes, such as personnel actions, payroll processing, and travel reimbursement. External customers are customers of the firm; agents for external customers, such as project designers; suppliers; and subcontractors and external agency representatives, such as building inspectors or construction-permitting authorities. Quality is a product of the culture of the company, and culture is the body of values and beliefs shared by all members of the firm. These values and beliefs are first expressed by company leaders and managers, and over time are adopted by the other members of the organization.

Total quality management is a process-focused management system that aims at continued increase in customer satisfaction at lower real cost. It works horizontally across functions and departments, and involves all employees, from top to bottom. TQM stresses continual learning and adaption to continual change as keys to organizational success. It requires the participation of each person in the company in identifying opportunities for improvement, determining performance requirements, defining the basis for measurement, and recommending methods of implementation. All individuals should have the opportunity to contribute, including all field offices and departments.

A major reason for implementing TQM within a construction firm is to enhance its competitive advantage in the market. The objective is to create a quality edge, as illustrated in Figure 9.3. First the quality gap must be eliminated by improving company processes to achieve customer expectations. Then the quality edge is created by exceeding customer expectations for service delivery.

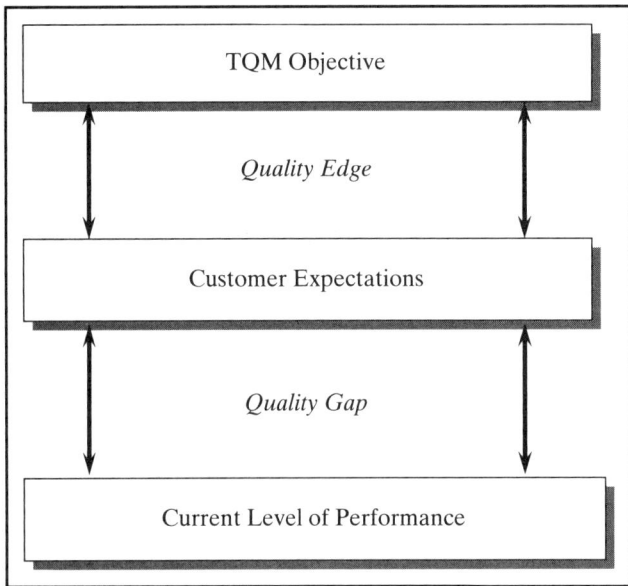

FIGURE 9.3 Use of TQM to Achieve Competitive Advantage

Quality Principles

The TQM process is built upon three fundamental principles—customer focus, process improvement, and empowerment and teamwork, as shown in Figure 9.4. Each is discussed below.

Customer focus: TQM is based on the concept that every business process used within the construction company has a customer and that the requirements and expectations of that customer must be met every time if the company as a whole plans to meet the requirements and expectations of its external customers. This requires a thorough understanding of customer requirements and a commitment to achieving them.

Process improvement: The concept of continuous process improvement was described in the Deming Quality Improvement Cycle shown in Figure 9.2. Each work

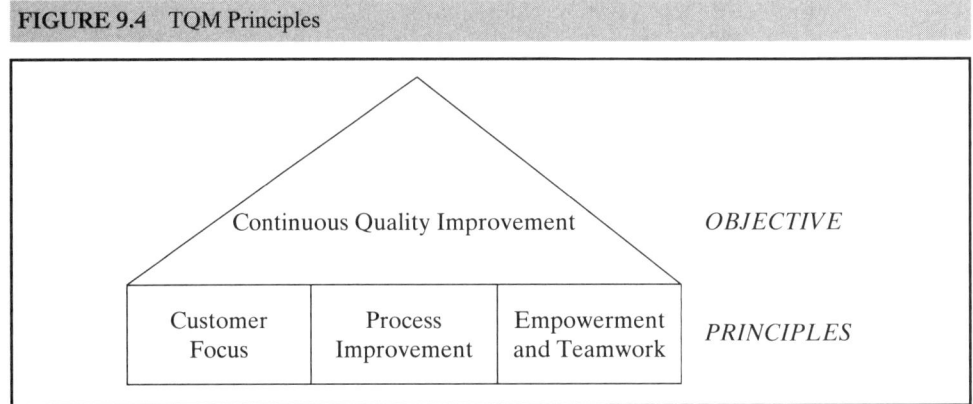

FIGURE 9.4 TQM Principles

process contains a series of steps to transform an input into an output. The goal of process improvement is to reengineer the process to ensure that the output meets all customer requirements.

Empowerment and teamwork: Empowerment means giving authority and responsibility to the people that are located where the work process occurs. It means granting authority to do whatever is necessary to satisfy customers and trusting employees to make the right choices without waiting for management approval. By empowering employees, companies drive decision making down to the lowest levels. Many firms have found that giving people throughout the organization the power to make decisions contributes greatly to providing quality products and services to their customers. Teamwork is essential to achieving continuous process improvement. A team approach helps break the mentality that employees should focus only on their individual responsibilities. Creating teams demonstrates top management's commitment to changing the organizational culture. Teaming reinforces the empowerment notion and provides various perspectives when identifying improvements to business processes. Teams established to analyze individual business processes often include suppliers and customers as well as service providers. This provides diverse perspectives to the team when it brainstorms ideas for process improvements.

Supporting Elements

There are six supporting elements that must exist within the company for TQM to be effective. These are

Vision and leadership: Senior management must lead the TQM effort by establishing a company vision that embraces quality as a core company value and creating a corporate culture that empowers employees to actively participate in improving the business processes of the firm.

Education and training: Quality is based on the understanding and the abilities of the employees of the company. Everyone within the organization needs to understand where the firm is going (as articulated in its company vision) and understand his or her role in accomplishing that vision. All employees must also understand the importance of quality and their roles in improving business processes to improve customer satisfaction.

Support structure: There needs to be an organizational structure within the company for managing the firm's TQM efforts. Managers at all levels must understand the importance of TQM initiatives and allow subordinates to participate in evaluating business processes.

Communications: Good communications are essential if TQM is to be effective. Employees need to understand the results obtained, and the teams involved in identifying process improvements need to be recognized. Good communications across functional departments, for example between field offices and the estimating department, are also needed to understand customer requirements and identify process improvements.

Recognition: Teams who successfully identify process improvements need to be recognized and rewarded. They should be rewarded collectively, not individually. The idea is to reinforce the importance of teamwork and the contribution of all team members.

Measurement: Process outputs must be measured and compared to customer requirements and industry standards. This is to quantitatively determine that individual business processes have been improved.

Process Improvement Model

The process improvement model used in TQM is illustrated in Figure 9.5. The first step is to identify the specific process to be evaluated. Next, the customer of the process, the owner of the process, and the supplier of input to the process are identified, as shown in Figure 9.6. Then customer requirements are established. This may be done by interviewing customers, by asking them to fill out questionnaires, or by conducting focus group sessions with them. The process is then examined in detail to determine all the

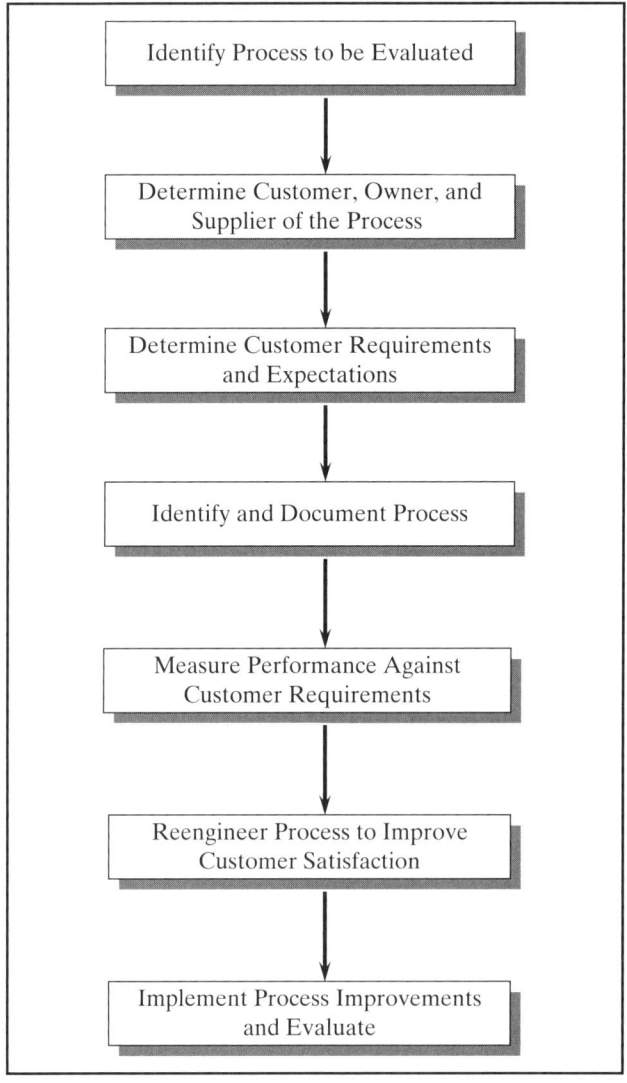

FIGURE 9.5 Process Improvement Model

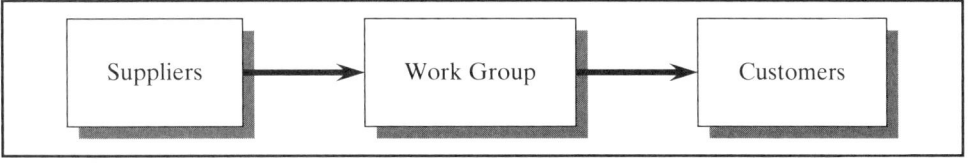

FIGURE 9.6 Work Process Model

steps in the process and the roles of all participants. Flowcharts are often used to map all the steps in an individual process. The process output is then compared to customer expectations to identify attributes of the output that do not meet customer requirements. Brainstorming sessions are then conducted to identify potential alternative techniques or procedures for accomplishing the process. These ideas are tested to evaluate their effectiveness in improving customer satisfaction. The most promising ideas are then used to reengineer the process.

Let's look at a construction example. The managers of Consolidated Construction Company have received numerous complaints from suppliers that they are not being paid timely. It takes two weeks from the submission of invoices to receipt of payment. A team was organized to evaluate the construction firm's payment procedures. The suppliers' expectations were to receive the correct payment within five business days. The team prepared a flowchart of the entire process from receipt of an invoice to issuance of a check. They then looked for alternative methods that could be used to improve customer (supplier) satisfaction. Rather than send a copy of the invoice to a project manager and then to an accounting technician, the team recommended that the invoice be scanned and sent electronically to all interested parties concurrently. All parties would be given one day in which to notify the finance department of any concerns that they had relative to the invoice. Once all parties concurred that payment should be made, the finance department would issue payment electronically to the supplier's bank. These revised procedures would allow the construction company to make payment within four business days, exceeding the suppliers' expectations. As a result, the suppliers may give the construction company more favorable terms on future purchases.

IMPLEMENTATION

Total quality management should not be thought of as a quick fix that can be implemented within a company quickly. It requires a culture change within the organization and total commitment from all managers and employees. It may take one to two years to plan TQM implementation, and another two to three years to implement the process. The basic steps in TQM implementation are shown in Figure 9.7.

For TQM to work within a company, quality must be specifically identified and adopted as one of the core company values. These company values play an important part in determining the attitudes of people at work. Positive attitudes enable people to work together effectively, while negative attitudes create barriers to effective communication. Employees must understand the tenets of TQM and be willing to collaborate freely for

184 CHAPTER 9 Total Quality Management

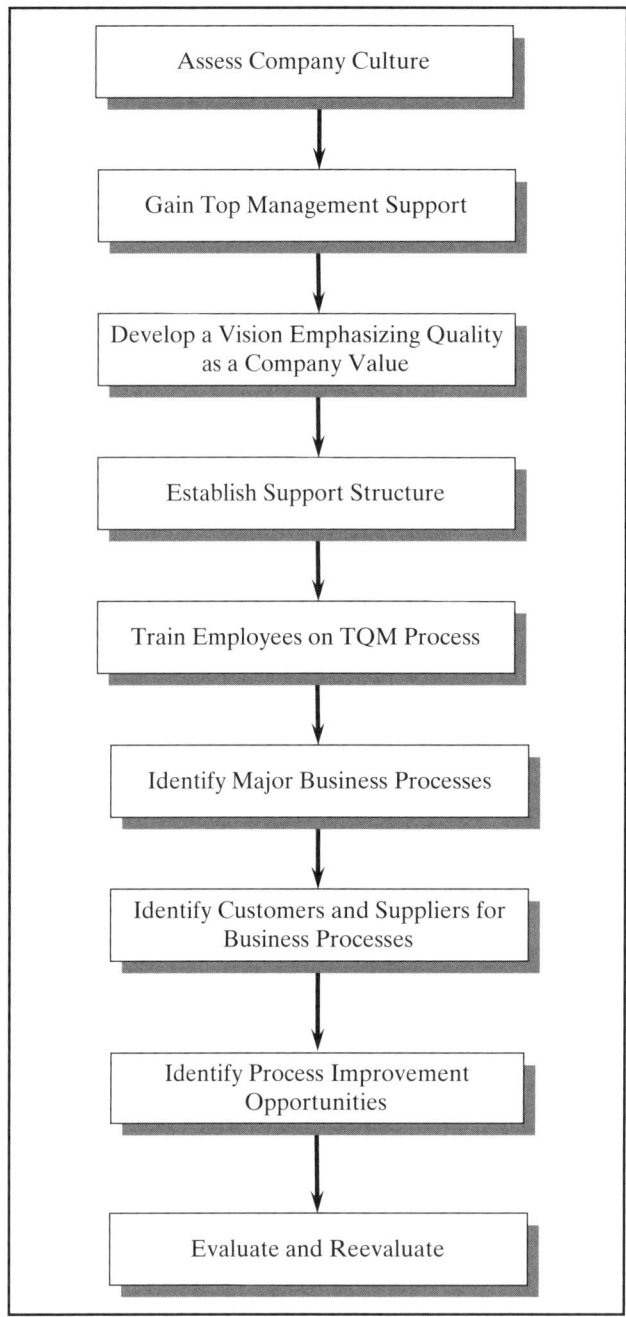

FIGURE 9.7 TQM Implementation Process

TQM to be effective within the company. These core company values are often stated in the company's mission statement and vision statement, as discussed in Chapter 5.

The first step in implementing TQM within a construction company is to assess the company culture. Does it employ centralized, hierarchical decision making? If so, TQM will not be effective, as it depends on empowered people able to make decisions regarding business processes. Managers must undertake initiatives to identify and change organizational conditions that make people feel powerless and increase employees' confidence that their efforts to make change will be successful. Empowerment and teamwork are essential if TQM is to be effective within the company.

Top management must understand what TQM is and how it works as well as be committed to the process. Otherwise, implementation will fail. Specialized training sessions for top management are needed before committing to implementation. Company leaders must agree that implementing TQM within the company is a good idea before proceeding. This is called the exploration and commitment phase of TQM implementation. Once the leadership has become committed, a TQM coordinator needs to be selected to coordinate all TQM implementation activities. TQM awareness training is scheduled for all employees to ensure that they understand why a TQM process is being implemented, what TQM is and how it works, and the potential improvements to be gained by its use. As a part of the awareness training, senior company leaders need to explain why a change is needed, why TQM has been selected as the vehicle for that change, and what the expected benefits to be realized by its implementation are.

The company strategic plan needs to be examined to ensure that quality is included in the company mission statement and in the company vision. If not, they need to be revised. Because of the time and resource commitment to TQM implementation, it needs to be included as one of the corporate strategies. An example of a quality corporate strategy might be

To build quality through teamwork to achieve total client satisfaction.

To manage TQM within the company, the support structure identified earlier, as the one with six essential supporting elements, needs to be established. The support structure used in larger construction companies is shown in Figure 9.8.

The **quality council** or steering committee is composed of top managers and is responsible for establishing total quality policy within the company and for guiding the implementation of TQM throughout the organization. The size of the council should

FIGURE 9.8 TQM Support Structure

be limited to seven to nine people for maximum efficiency. The quality council may meet frequently when the effort is getting started, but usually meets only monthly or quarterly once implementation is under way. Meetings typically occur only quarterly once the company has moved from implementation into sustainment. The group makes key decisions regarding how quality should be measured and what approaches should be used to improve quality. They need to identify any barrier that could impede implementation of TQM and devise methods for eliminating those barriers. The quality council also periodically reviews the status of TQM within the company and makes adjustments to ensure customer satisfaction and continuous improvement. In general, the quality council has overall responsibility for the progress and success of the TQM effort. The council needs to ensure the company recognition program's transition from individual recognition to team recognition. This will reinforce the importance of teamwork and employee participation on teams.

The **quality management board** is composed of senior managers, and they are responsible for managing quality improvement activities within the company. The board selects the processes to be evaluated, appoints the process action teams, and reviews and approves the teams' process improvement recommendations. The size of the quality management board should also be limited to seven to nine people. In small companies, the functions of the quality management board and the quality council are often combined into a single quality management steering committee. The quality management board develops a flowchart of all the company business processes and surveys external and internal customers, regarding their perceptions of the quality of services provided. The flowchart should identify what functions are performed, who is responsible for them, who accomplishes them, and how they are interrelated. The initial processes selected for analysis should have high visibility to customers and have a high potential for improvement. To reinforce the value of TQM, managers want the initial process action teams to be successful and to publicize their accomplishments.

The **process action teams** perform the process analysis, using the process improvement model shown in Figure 9.5. These teams should be composed of people possessing the best knowledge regarding the process to be evaluated. The key to making TQM work is empowering the people closest to each process with the authority to make improvements. This capitalizes on the knowledge and experience of those individuals in the best position to identify process improvements. Individuals assigned to a process action team should include individuals who work with the process, individuals who provide input to the process, and customers of the process. This ensures that all aspects of the process are understood by the team and that all perspectives are considered during team deliberations. Team training is needed by the process action teams to ensure that they understand the TQM philosophy, the process improvement model, the brainstorming techniques, the importance of regarding differences of opinion as opportunities for mutual growth, and the need to identify collective ideas regarding process improvements. Team-building exercises may be needed to develop cohesive teams. The size of the process action team should be five to seven people for maximum efficiency.

The number of process action teams needs to be limited to control the cost of TQM and to not overwhelm the quality management board. Initial implementation should be limited to two or three teams. Small companies may not establish more than five teams, and larger companies should limit the number of teams to about ten.

Membership on a process action team is not a permanent assignment. Once process improvements have been implemented, the team may be eliminated and a new one established to analyze another process. Process action teams should be given goals and timelines. They may meet for one or two hours per week for one to six months, and they should report progress to the quality management board monthly. Senior managers should ensure that self-motivated, inquisitive individuals are assigned to the process action teams and that their supervisors understand the importance of their services as team members.

PROCESS ANALYSIS

What type of processes should be selected for analysis? Company leaders should select those that have the greatest impact on both external and internal customers. The basic business processes of a construction firm, generally, can be categorized as administrative, business development, project management, logistical, or construction processes. Some examples are shown below:

Administrative processes

- Internal communication
- Invoice processing
- Overhead budgeting
- Payroll
- Personnel actions

Business development processes

- Current customer relationship enhancement
- Marketing media
- Potential customer tracking
- Public relations

Project management processes

- Contract administration
- Cost estimating
- Lien release management
- Operation and maintenance document management
- Progress payment request submission
- Project scheduling
- Request for Information management
- Subcontract management
- Subcontractor procurement
- Submittal management

Logistical processes

- Equipment management (purchase, rent, lease, or sell)
- Fuel management

- Material procurement
- Material storage
- Office space management

Construction processes

- Equipment management (utilization)
- Quality control
- Safety management
- Subcontractor coordination

Once a process to be analyzed has been selected, a process action team is organized that contains individuals representing those people who provide input to the process, those who are recipients of the output of the process, and those who participate in performing the process. The size of the team should be limited to five to seven people. Once organized, the team members need to get to know one another and establish procedures for conducting their assessment, brainstorming alternative approaches, and selecting recommended improvements. Now it is time to conduct the analysis of the assigned process or processes. A flowchart of a procedure that can be used for process analysis and improvement is shown in Figure 9.9. The basic idea is to determine the customers and suppliers of the process, the expectations of the customers, the components of the process, and the characteristics of the current output. Based on the analysis, alternative approaches to process performance are developed and tested. Those approaches that appear to have the greatest potential for increased customer satisfaction are selected, tested, and then implemented. The results are evaluated to determine the effect of the identified process improvements.

SUSTAINMENT

Sustaining TQM within a company requires total commitment from the senior leadership. They must continually drive decision making and problem resolution to the lowest practicable level and encourage a team approach to process improvement. TQM training will need to be conducted every year to ensure that new employees understand the company's commitment to quality and their roles as members of the company team. Team training will be needed as new process action teams are organized, to ensure that they have the skills and perspective to work as teams and to identify process improvements. TQM organizations tend to be learning organizations and depend on employees becoming increasingly competent and creative in their quest for quality. Teams are the primary means for planning and problem solving. This requires developing trusting relationships among all members of the company. Trust and empowerment are critical if TQM is to be effective within the company. The important relationship between empowerment and quality is illustrated in Figure 9.10.

One of the issues to be faced in sustaining TQM is the establishment of goals for process improvement. Many companies are using a benchmarking approach, which involves identifying the best practices within the industry. The steps in benchmarking are determining the functions to be benchmarked, identifying the performance indicators to be measured, and measuring the performance of the best-performing companies. Benchmarking is basically the search for the best practices, which are often difficult to

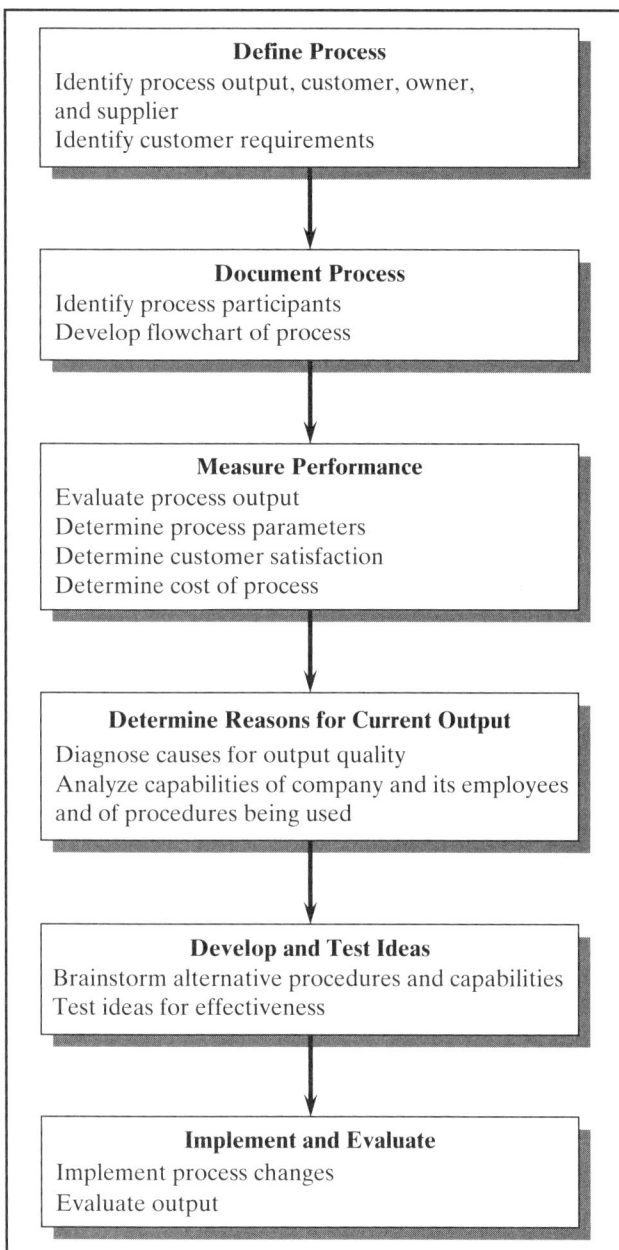

FIGURE 9.9 Process Analysis

quantify. These benchmarks then become the improvement goals for the respective processes. The gaps between the company's performance and the benchmarks define the challenges for the process action teams in identifying methods for process improvement. The differences between the company's performance in some areas and the benchmarks should be a topic for discussion by the quality council at its quarterly meeting.

Implementing TQM within a construction company also involves evaluating field procedures and processes on construction projects. These are more complicated to

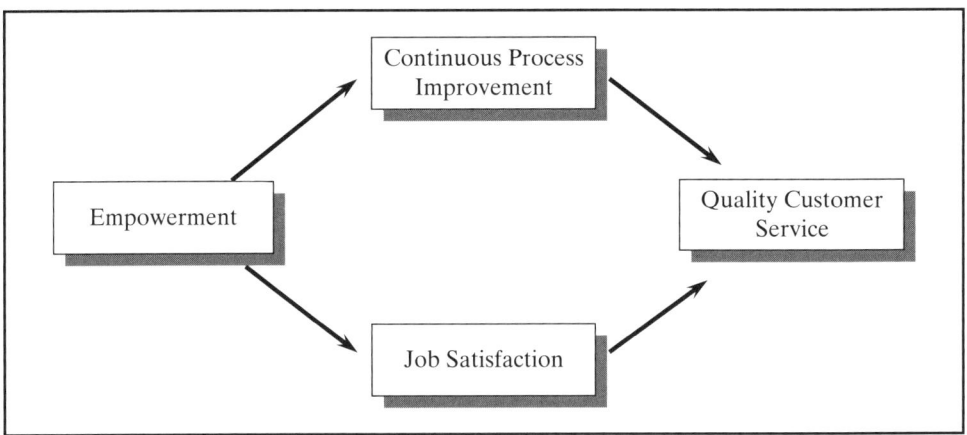

FIGURE 9.10 How Empowerment Leads to Quality

analyze because they typically involve individuals from many organizations, as indicated in Figure 9.11. A single steering committee should be organized with senior representatives from each of the major participating organizations. Two or three process action teams should be established composed of members from all affected organizations. Processes to be analyzed may include procedures for processing requests for information, shop drawing submittal process, contract change order procedures, scheduling subcontractors, material procurement procedures, safety management procedures, and dispute resolution procedures. Specific construction activities, such as concrete form construction, could also be analyzed by process action teams to identify more efficient techniques for accomplishing them. The objective may be to identify methods for improving workmanship, reducing material waste, improving productivity, and improving safety.

FIGURE 9.11 Construction Project Participants

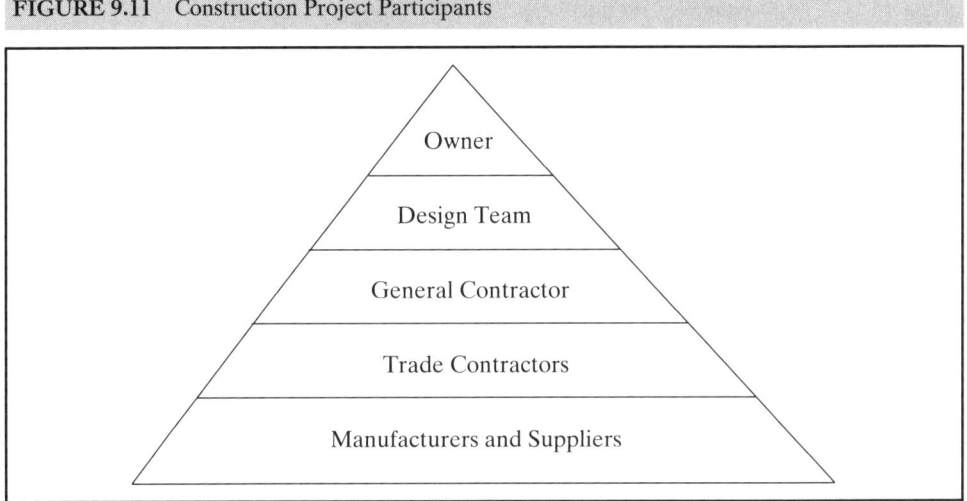

Many construction companies that have adopted TQM have found that the return on their investment far exceeded the cost of implementation. Most had to change their business culture by empowering their employees to identify business process improvements. Supervisors must also be educated to the need for allowing subordinates to participate in process action teams. Successful implementation of TQM should allow the company to realize higher productivity, better employee relations, greater customer satisfaction, increased market share, and improved profitability.

EFFECTIVENESS ASSESSMENT

To assess the effectiveness of TQM within the construction company, periodic surveys should be taken of both internal and external customers to ascertain their satisfaction with services being received. Analysis of survey results will indicate which processes need further improvement. Review of trends in survey responses will provide a measure of the effectiveness of TQM in improving customer satisfaction and in reducing the cost of doing business. Survey results should be reviewed by the quality council at least quarterly to provide them with an evaluation of the implementation of TQM within the company.

The United States Government recognized the need to focus on quality as a major factor for industries to be competitive in the global economy. In 1987, Congress authorized an award for companies that are models in demonstrating quality products and services. This award, the Malcolm Baldridge National Quality Award, is named after Malcolm Baldridge, a former secretary of commerce. The criteria used in evaluating companies for the award are shown in Figure 9.12. Within each category, the scoring system is based on three factors—approach, deployment, and results. *Approach* refers to the methods used to accomplish the objective. *Deployment* refers to the extent to which the approach has been deployed within the company. *Results* refers to the outcomes and effects achieved. These same criteria could be used by a construction company in evaluating the effectiveness of its TQM program.

Summary

Total quality management is a proactive management philosophy that focuses on continuous process improvement and customer satisfaction. Introduced into the United States in the 1980s, it has been widely adopted by manufacturing and service companies. The construction industry has been slow to adopt TQM, but a number of

FIGURE 9.12 Malcolm Baldridge Award Criteria

- Leadership
- Information and Analysis
- Strategic Quality Planning
- Human Resources Development and Management
- Management of Process Quality
- Quality and Operational Results
- Customer Focus and Satisfaction

construction firms have successfully implemented the process. Meeting customers' requirements applies to both internal and external customers. TQM requires a total commitment to continuous process improvement to achieve customer satisfaction. It requires a team approach to problem solving and empowered employees willing to make decisions. To manage TQM within a company, an organizational support structure needs to be established. A typical organization consists of a quality council, a quality management board, and several process action teams. In small companies, the quality council and the quality management board functions may be combined into a single committee. The number of process action teams needs to be limited to control the cost of TQM and to not overwhelm the management structure. Sustaining TQM within a company requires total commitment from the senior leadership and changing the rewards system from individual-based to team-based. Benchmarking may be used to establish process improvement goals for the company. Successful implementation of TQM should allow the company to realize higher productivity, better employee morale, greater customer satisfaction, and improved profitability.

Review Questions

1. Describe how Deming's plan-do-check-act cycle might be applied in improving a construction company's procedures for scheduling subcontractors' work on a construction project.
2. What are the three fundamental principles of TQM? Why are they critical to the implementation of the process?
3. What types of education and training are needed during the TQM implementation phase?
4. What are the basic steps in the TQM process improvement model?
5. What are the roles and responsibilities of the quality council?
6. What are the roles and responsibilities of the quality management board?
7. What are the roles and responsibilities of a process action team?
8. What is benchmarking and how is it done?

Exercises

1. If you were the president of a medium-size general construction firm, who would you assign to the quality council? Who would you assign to the quality management board?
2. Who would you assign to a process action team tasked with evaluating the cost-estimating process used in a construction firm?
3. Develop a flowchart for the submittal process used on a typical construction project. Who is the customer, and how would you determine the customer's requirements?
4. Develop a flowchart for the process for submission of monthly progress payment requests on a typical construction project. Who is the customer, and how would you determine the customer's requirements?
5. Identify all the processes that take place between a general contractor and a subcontractor on a typical construction project. Identify the owner and the customer for each process.

Sources of Additional Information

Chase, Gerald W. *Implementing TQM in a Construction Company*, Washington, D.C.: Associated General Contractors of America, 1993.

Deming, W. Edwards. *Out of Crisis*, Cambridge, Mass.: MIT Center for Advanced Engineering Study, 1986.

Implementing TQM in Engineering and Construction, Special Publication 31-1, Austin, Tex.: Construction Industry Institute, 1994.

Schmidt, Warren H., and Jerome P. Finnigan. *TQManager: A Practical Guide for Managing in a Total Quality Organization*, San Francisco: Jossey-Bass Publishers, 1993.

Summers, Donna C. S. *Quality Management: Creating and Sustaining Organizational Effectiveness*, Upper Saddle River, N.J.: Prentice-Hall, 2005.

Tenner, Arthur R., and Irving J. DeToro. *Total Quality Management: Three Steps to Continuous Improvement*, Reading, Mass.: Addison-Wesley Publishing Company, 1992.

Walton, Mary. *Deming Management at Work*, New York: G. P. Putnam and Sons, 1991.

Appendix A

Strategic Plan for Pacific Constructors

Background: Pacific Constructors is a medium-sized construction company that specializes in mixed-use, hospitality, retail, and medium-sized commercial projects. Historically, the company has only pursued projects with private clients.

Strategic Assessment

External Analysis

General Business Environment

Economic environment. Inflation of material costs will continue at a moderate rate over the next five years and interest rates will remain relatively stable. The availability of project financing will enable current and potential clients to continue to invest in new facilities. The business climate will be supportive of continued facility investments.

Demographic environment. The construction industry will continue to have an aging workforce, and recruiting new employees will be a challenge. The population in our market area will continue to grow as new businesses are established, and existing businesses expand. A high percentage of new industry employees will be immigrants for whom English will not be their native language.

Political/legal environment. Growth management regulations will continue to emphasize increased population density in metropolitan areas, and municipalities will implement zoning regulations requiring that more amenities be provided with new projects. Environmental regulation will be increased, requiring reduction of waste from construction projects. Material reuse policies will be strengthened, and sustainable construction techniques will be mandated.

Technological environment. The use of building information models and other visualization tools will become the standard tools for planning and designing projects. Information technologies, recycled materials, and composite materials will be more extensively used in construction. Use of Internet-based collaborative and project management tools will be more extensively used.

Sociocultural environment. The number of people looking for part-time employment and flexible working hours will increase. Attracting young men and women to the construction industry will be a challenge. Increased employment of individuals for whom English is not their native language will present increased communication challenges on construction projects.

Specific Industry Environment

Customer analysis. The demand for construction services among current and potential clients will decrease over the next five years. Clients will be more selective in the selection of their construction agents. Project quality and challenging schedules will continue to be the clients' primary goals. Cost must be reasonable, but clients understand that they may need to pay premium for high-quality service. Personal relationships will continue to be critical to clients in the selection of construction contractors.

Subcontractor analysis. Demand for quality subcontractors will continue to exceed the supply in the Northwest. Subcontractors will request higher fees on high-risk projects or not submit proposals.

The number of subcontractors submitting proposals on individual projects will decline as subcontractors continue to have difficulty recruiting craft labor.

Supplier analysis. Demand for construction materials will continue to challenge suppliers. Material requirements for Asia will continue to place pressure on U.S. suppliers, and prices will increase at a moderate rate. The moderation of demand for construction and increased reuse of materials will decrease the demand for new materials.

Competitor analysis. The number of competitors is expected to remain relatively constant over the next five years. As the demand for construction services declines, competition will intensify; and clients will be more selective in hiring builders for their projects.

Internal Assessment

Strengths

- The company has a highly motivated team of employees, who possess excellent technical skills and provide exceptional customer service.
- There has been little employee turnover during the past five years, and 90 percent of the employees hired during the period are still employed by the company.
- About 80 percent of the company's business during the past three years has been with repeat clients.
- Working capital, profitability, business volume, owners' equity, and liquidity have increased at a steady rate over the past five years.
- The company has upgraded its information management systems over the past three years and currently has state-of-the-art equipment and software.
- The company has identified replacements for three executives who will be retiring in the next five years and is actively developing their abilities to assume their new responsibilities.

Weaknesses

- The company needs to attract entry-level project management personnel to create additional project management teams.
- The company needs to attract experienced superintendents to enable the undertaking of additional projects.
- The company needs to implement a more efficient system for managing its overhead costs.

Strategic Direction

Mission Statement

Pacific Constructors is an industry leader known for

- Innovative projects.
- Developing enduring relationships with clients, designers, subcontractors, and suppliers.
- Investing in our employees to ensure that they have the needed skills.

Vision

To become the builder of choice for clients looking for the best value in their service provider.

Strategic Objectives

- Increase the business volume by 5 percent per year over the next five years.
- Increase profitability and owners' equity by 5 percent per year over the next five years.
- Increase the percentage of work done for new clients to 25 percent per year to diversify our company's client base.
- Maintain good relationships with existing clients and cultivate good relationships with new clients.
- Recruit five new project engineers and three new experienced superintendents per year over the next five years to increase company project management capabilities.
- Improve management of company overhead budget.

APPENDIX A Strategic Plan for Pacific Constructors

Strategies to Accomplish Objectives

- Cultivate enduring relationships with design firms who can provide contacts with potential clients.
- Maintain personal relationships with current clients to understand their needs and procurement practices.
- Follow-up with each client about six months after project completion.
- Ensure that fee proposals provide the desired financial return.
- Participate in community activities to enhance company name recognition among potential clients.
- Manage construction projects to complete them within negotiated costs and schedules.
- Participate in university career fairs and offer summer internships to prospective new project engineers.
- Identify potential future superintendents among the company foremen and assign them to development programs to enhance their technical and leadership skills.
- Acquire state-of-the-art financial management software to manage the company overhead budget.

Action Plans to Execute Strategies

- *Cultivate enduring relationships with design firms who can provide contacts with potential clients.*
  - Meet with design firm principals to discuss current and anticipated projects that fit within our company's scope of interest.
    - Responsible person: Director of Business Development
    - Resources allocated: $10,000 per year
    - Date to be accomplished: Ongoing
  - Invite design firm principals to golf outings or sports venues to cultivate personal relationships.
    - Responsible person: Director of Business Development
    - Resources allocated: $30,000 per year
    - Date to be accomplished: Ongoing

- *Maintain personal relationships with current clients to understand their needs and procurement practices.*
  - Invite counterparts from client firms to golf outings, sport venues, or other social activities to cultivate personal relationships.
    - Responsible person: Director of Construction Operations
    - Resources allocated: $55,000 per year
    - Date to be accomplished: Ongoing

- *Follow-up with each client about six months after project completion.*
  - Develop questionnaire and send to clients six months after completion of each project. Follow-up with clients on any issues that indicate dissatisfaction.
    - Responsible person: Director of Construction Operations
    - Resources allocated: $105,000 per year
    - Date to be accomplished: Ongoing; initial questionnaire within three months

- *Ensure that fee proposals provide the desired financial return.*
  - Develop realistic estimates for costs to manage and execute projects.
    - Responsible person: Director of Construction Operations
    - Resources allocated: $100,000 per year
    - Date to be accomplished: Ongoing
  - Establish realistic company overhead budget and manage overhead costs to remain within the budget.
    - Responsible person: Chief Financial Officer
    - Resources allocated: $30,000 per year
    - Date to be accomplished: Ongoing; initial budget within three months

- *Participate in community activities to enhance company name recognition among potential clients.*
  - Select community events in which company employees should participate.
    - Responsible person: Director of Business Development

Resources allocated: $5,000 per year

Date to be accomplished: Ongoing; initial list within six months

- Select employees to participate in selected community events.

    Responsible person: Vice President

    Resources allocated: $45,000 per year

    Date to be accomplished: Ongoing

- *Manage construction projects to complete them within negotiated costs and schedules.*
    - Properly staff the project management teams for all projects and provide oversight of teams.

        Responsible person: Director of Construction Operations

        Resources allocated: $1.5 million per year

        Date to be accomplished: Ongoing

- *Participate in university career fairs and offer summer internships to prospective new project engineers.*
    - Develop company recruiting materials and participate in career fairs at universities offering degrees in construction management.

        Responsible person: Human Resources Manager

        Resources allocated: $15,000 per year

        Date to be accomplished: Ongoing; recruiting materials within four months

    - Develop and implement a summer internship program for university students.

        Responsible person: Director of Construction Operations

        Resources allocated: $20,000 per year

        Date to be accomplished: Ongoing; initial program within six months

- *Identify potential future superintendents among the company foremen and assign them to development programs to enhance their technical and leadership skills.*
    - Identify good candidates for promotion from superintendents and project managers, create individual development plans for them, and monitor their progress.

        Responsible person: Director of Construction Operations

        Resources allocated: $75,000 per year

        Date to be accomplished: Ongoing

- *Acquire state-of-the-art financial management software to manage the company overhead budget.*
    - Procure and implement a financial management system to manage the company overhead budget.

        Responsible person: Chief Financial Officer

        Resources allocated: $100,000

        Date to be accomplished: Ongoing; initial list within six months

Financial Plan

- Retain 60 percent of net profits each year to increase owners' equity and reduce the need to borrow funds to finance negative cash flow.

    Responsible person: Chief Financial Officer

    Resources allocated: None needed

    Date to be accomplished: Ongoing

- Obtain a $1-million line of credit each year that can be used to cover unexpected expenses.

    Responsible person: Chief Financial Officer

    Resources allocated: $8,000 per year

    Date to be accomplished: January 1 of each year

- Submit all requests for payment invoices in a timely manner, and take all discounts offered by suppliers for early payment of material invoices.

    Responsible person: Director of Construction Operations

    Resources allocated: None needed

    Date to be accomplished: Ongoing

- Continue to finance equipment acquisitions with loans from equipment finance companies.
  - Responsible person: Equipment Manager
  - Resources allocated: None needed
  - Date to be accomplished: Ongoing

Business Development Plan

- Create and maintain a "target list" of desired prospective clients and their anticipated projects.
  - Responsible person: Director of Business Development
  - Resources allocated: $12,000 per year
  - Date to be accomplished: Ongoing; initial list within six months
- Create and maintain a "target list" of architects who can provide design services for current and desired prospective clients.
  - Responsible person: Director of Business Development
  - Resources allocated: $8,000 per year
  - Date to be accomplished: Ongoing; initial list within six months
- Create and maintain a list of projects that our major competitors receive from current and desired prospective clients.
  - Responsible person: Director of Business Development
  - Resources allocated: $20,000 per year
  - Date to be accomplished: Ongoing; initial list within six months
- Assign project managers to cultivate and maintain close personal relationships with counterparts employed by major clients.
  - Responsible person: Director of Construction Operations
  - Resources allocated: $55,000 per year
  - Date to be accomplished: June 20XX
- Participate in community charity events to enhance company name recognition among general public.
  - Responsible person: Director of Business Development to identify events, and Vice President to identify company participants
  - Resources allocated: $50,000 per year
  - Date to be accomplished: Ongoing
- Develop a consistent company "brand" to be used on all company correspondence, signs, and vehicles.
  - Responsible person: Director of Business Development
  - Resources allocated: $20,000 per year
  - Date to be accomplished: September 20XX
- Create a new company brochure that can be tailored to focus on individual prospective clients.
  - Responsible person: Director of Business Development
  - Resources allocated: $10,000 per year
  - Date to be accomplished: June 20XX
- Issue press releases to announce the completion of all company projects and awards to employees.
  - Responsible person: Director of Business Development
  - Resources allocated: $10,000 per year
  - Date to be accomplished: Ongoing

Resource Plan

Capital Assets

- Replace company vehicles after 100,000 miles of use.
- Replace construction equipment after 18,000 hours of use.
- Lease or rent equipment not routinely used by company employees on projects.

Human Resources

- Invest 3 percent of annual payroll budget on employee training and development.
- Actively recruit at universities and technical schools to attract new talent.
- Ensure that employee development plans are developed for all employees.
- Expand the company wellness program to include covering 50 percent of employee membership fees in health clubs and fitness centers.

- Recruit sufficient new employees to allow the company to undertake the 5 percent increase in business volume per year.

Information Management

- Continue to upgrade information management security systems.
- Invest in hardware and software upgrades each year.
- Conduct literacy training on the use of new systems and to new employees.

Contingency Plan

Monitor the business environment on a quarterly basis to determine if it evolves as envisioned in the External Analysis. If not, make adjustments in capital investment and new hiring to avoid costs that may not be covered by revenues. Fees may be reduced to remain competitive, but each project must be priced to generate a profit.

Appendix B

CASE STUDY
Cascade Builders Annual Report

Company Description

Overview. Cascade Builders, referred to as the Company in this report, operates as a diversified builder of medium-priced, single-family homes for use as primary residences with operations in major metropolitan markets of western Washington. The Company is a privately owned corporation organized under the laws of the State of Washington. For the year ending December 31, 20XX, approximately 23 percent, 16 percent, and 20 percent of the Company's home deliveries were in Snohomish, Pierce, and King Counties, respectively. During the year, an additional 18 percent, 15 percent, and 8 percent of the total constructed homes were in Kitsap, Mason, and Thurston Counties. Also during the year, 100 percent of the Company's sales and revenues were derived from home-building activities.

Employment. On December 31, 20XX, the Company had 28 employees. During the previous five years, the Company has not directly experienced a work stoppage in its operation caused by labor disputes. Delays during that time due to strikes by construction unions against subcontractors retained by the Company have not had a significant adverse effect on the Company's home-building operations.

Properties. In addition to real estate held for development and sale, which is either owned or under option to be purchased by the Company, the Company leases approximately 1.8 acres of land in Issaquah, Washington, under leases expiring in 2020, on which the Company's executive office is located.

Company History

Overview. The Company began its home-building operations in 1980 with a single tract of land in King County, Washington, and currently conducts home-building activities in western Washington in six different counties.

From the beginning, the Company designed, built, and sold single-family homes. Initially they were designed to appeal to entry-level and first move-up homebuyers. Until the end of the 1980s, the Company's market was primarily in King County. During the 1990s, the Company expanded its operations to include Pierce and Snohomish Counties, following the shift of new housing out of King County and into neighboring counties. Today the Company services a diversified customer base in the western Washington metropolitan areas where much new residential construction prevails.

Typical Product Description. Substantially all of the Company's homes sold have been single-family detached dwellings, although during the 1990s about 7 percent were townhouses or condominiums attached in varying configurations of two, three, four or five dwellings. The Company's homes have been designed to suit the particular area of the region they are located in and have been available in a variety of models, exterior styles, and materials, depending upon local preferences. Homes built by the Company have been targeted for occupancy as primary residences. Homes built by the Company typically range in size from approximately 2,000 to 3,000 square feet and typically include three or four bedrooms, two or three baths, a living room, kitchen, dining room, family room, and a two- or three-car garage. As mentioned above, the Company has also built single-family attached dwellings ranging from 1,100 to 5,500 square feet. Gas fireplaces and built-in appliances have usually been included in all residential designs.

Company Organization and Responsibilities

Each home-building project is run by a local project manager. Decisions regarding selection of parcels of land to purchase and develop are made in conjunction with the officers of the Company, and thereafter, each manager conducts the operations of the project autonomously as a separate profit center. The managers, often assigned to more than one project, report to the Company's Vice President for Construction, who oversees their development. The Vice President in turn reports to the Company's President. The Company employs a Field Superintendent at each project site to supervise actual construction. Finally, each geographic area has one customer service and marketing representative assigned to projects being constructed in that area.

The Company essentially functions as a general contractor with its supervisory employees coordinating work on the project. The services of independent architectural, engineering, and other consulting firms are engaged to assist in project planning, and subcontractors are employed to perform all of the physical development and construction work on the project. The Company does not have long-term contractual commitments with any of its subcontractors, consultants, or suppliers of materials. However, because of its market presence and long-term relationships, the Company has generally been able to obtain sufficient materials and commitments from subcontractors during times of market shortages and downturns. These types of agreements are generally entered into on an increment-by-increment basis at a fixed price after competitive bidding. The Company believes that the low fixed labor expense resulting from conducting its home-building operations in this manner has been instrumental in enabling it to retain the necessary flexibility to react to increases or decreases in demand for housing. Although construction time for the Company's homes varies from project to project, depending on the time of year, local labor situations, certain governmental approval processes, availability of materials and supplies, and other factors, the Company can typically complete the construction of each house in approximately six to eight months.

Company officers complete a purchase of land only when they can reasonably project the beginning of construction within one year. Closing of the land purchase is made contingent upon satisfaction of conditions relating to the property and to the Company's being able to obtain all approvals from governmental agencies within a year. The Company typically uses unsecured financing in the form of bank debt and other unsecured debt to fund land acquisitions. The Company usually acquires unimproved or improved land zoned for residential use that appears suitable for the construction of 20 to 100 homes in increments of 5 to 20 homes. The number of homes built in the first increment of a project is based upon the Company's market studies. Development work on a project includes obtaining all necessary zoning, environmental and other regulatory approvals, and constructing roads, sewer and drainage systems, recreational facilities, and other improvements.

The following chart illustrates the Company's organization (Figure B.1):

Current Company Trends

Results of Recent Operations. During the year ending December 31, 20XX, the Company delivered 73 new homes at an average selling price of $309,407 compared to 82 new homes at an average selling price of $279,073 in the previous year. The average selling price of the Company's homes was impacted by product mix, geographic mix, and changing prices on homes sold. The yearly increases in the average selling prices were primarily due to the increased deliveries of homes in projects located in Snohomish and Pierce Counties and to an increasing cost of land and building materials, especially wood products. Cost of sales increased substantially from the previous year. The increase was primarily due to the increase in marketing and sales activity as well as the cost of materials.

Business Risks. The development, construction, and sale of homes is subject locally to various risks, including the continued availability of suitable undeveloped land at reasonable prices and adverse local market conditions resulting from changes in economic conditions or competitive overbuilding. Other risks include changing governmental regulations; increases in prevailing interest rates; the level of real estate taxes; the cost of materials and labor; the availability of construction financing, and home mortgage financing attractive to the purchasers of the

FIGURE B.1 Organization Chart

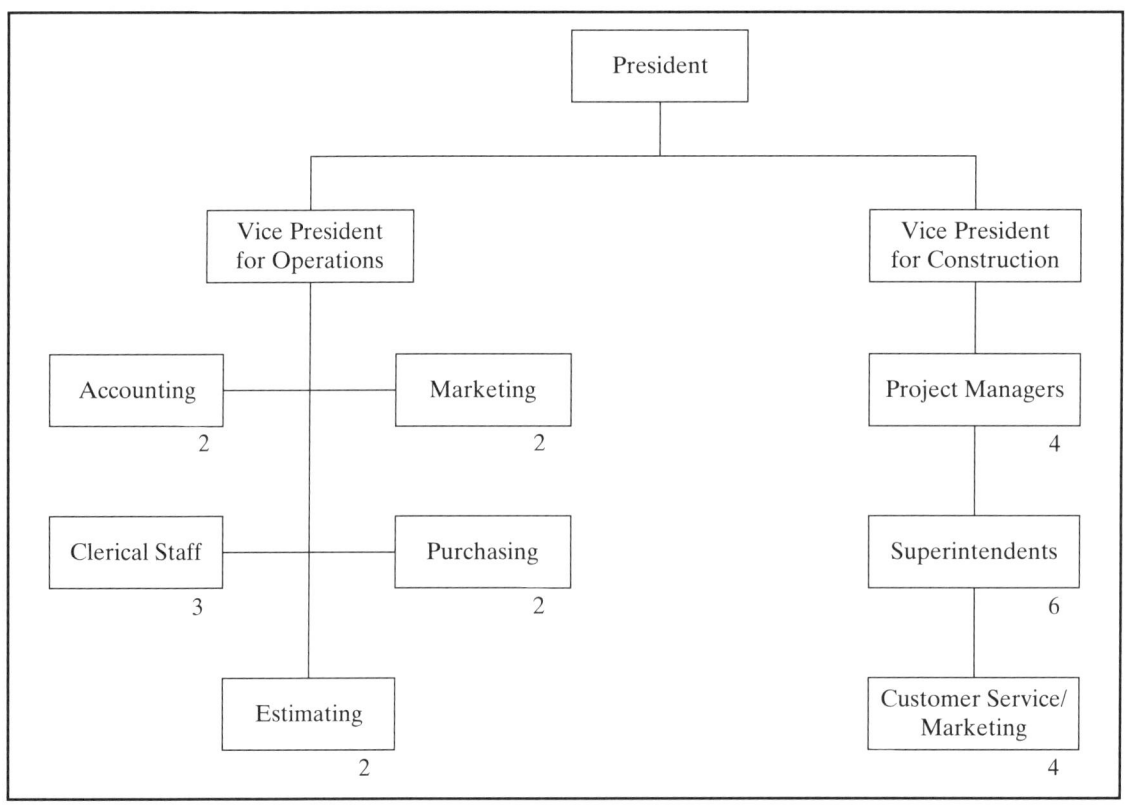

Company's homes—and inclement weather. The Company's business is affected by the general economic conditions in the markets it serves in Washington and by the level of mortgage interest rates and consumer confidence in those regions.

Regulation. The housing industry is subject to increasing environmental, building, zoning, and real estate regulations by various Federal, state, county, and city authorities. In developing a project, the Company must obtain the approval of numerous governmental authorities regulating such matters as permitted land uses and levels of density, the installation of utility services, such as water and waste disposal, and the dedication of acreage for open space, parks, schools, and other community purposes. Although the Company believes that it has acquired a sufficient number of lots to provide for its home construction activities for the near term, no assurances can be given that the Company will be able to sell the homes it produces on a profitable basis.

Financing. Home purchase financing from local lending institutions generally averages 80 percent of the purchase price of the homes. During periods of high mortgage rates or difficult economic times, the Company may assist its homebuyers by "buying-down" the interest rates on mortgage loans or subsidizing all or a part of the homebuyers' upfront financing fees. The amounts of such "buy-downs" or subsidies are dependent upon prevailing market conditions and interest rate levels.

Current Marketing Activities

The Company makes extensive use of advertisements in local newspapers, illustrated brochures, direct mail, and the placement of strategically located sign boards in the immediate areas of its developments. The Company's homes are generally sold by its own staff of sales personnel. Between one and three furnished and landscaped model homes are usually maintained at each project site. The Company believes model homes play an important role in its marketing efforts. Consequently, the Company expends a significant effort in creating an attractive atmosphere at its model homes. Interior decorations are undertaken by independent design specialists. The purchase of furniture, fixtures, and fittings is coordinated to ensure that manufacturers' bulk discounts are utilized to the greatest extent. Changes in house design are generally not permitted, but homebuyers are afforded the opportunity to select, at additional cost, various optional amenities such as air conditioning, pre-wiring options, upgraded carpet quality, varied interior and exterior color schemes, additional appliances, and occasionally some room configurations.

The Company sells its homes through commissioned employees, who typically work from the sales offices located at the model homes used in each subdivision. Company personnel are available to further assist prospective homebuyers by encouraging them to visit other Company projects based on the customers' needs. The Company's homes are typically sold during construction, using sales contracts which are usually accompanied by a small cash deposit, although during the last several years an increasing number of the Company's homes have been sold after completion of construction. In some cases, purchasers are permitted to cancel their contracts if they are unable to sell their existing homes or fail to qualify for financing. Sales are recorded after construction is completed, required down payments are received, and title passes.

Description of Competition

The development and sale of residential properties is highly competitive and fragmented. The Company competes for residential sales on the basis of a number of interrelated factors—including location, reputation, amenities, design, quality, and price—with numerous large and small homebuilders, including some homebuilders with nationwide operations and greater financial resources and/or lower costs than the Company. The Company also competes for residential sales with individual resales of existing homes, available rental housing and, to a lesser extent, resales of condominiums. The Company believes that it compares favorably to other builders in the markets in which it operates. This is primarily due to its experience within its geographic markets, which allows it to vary its product offerings to reflect changing market conditions; its responsiveness to market conditions, enabling it to capitalize on the opportunities for land acquisitions in desirable locations; and its reputation for service and quality.

Description of Major Customers

The Company has focused on move-up homebuyers because it believes they represent the largest segment of the homebuilding market. While attempting to remain responsive to market opportunities within the industry, the Company in recent years has focused its business primarily on move-up homes priced at the middle of the market and to target second- and third-time homebuyers. First and second move-up homes (as opposed to entry-level homes) usually attract a wide variety of homebuyers as they progress in income and family size. Although some of the Company's move-up homes are priced at the upper end of the market and the Company offers a selection of optional amenities, the Company does not build "custom homes," and its prices for first move-up homes, generally, are below the prices of custom homes in most project regions. The Company attempts to maximize efficiency and keep prices down by using standardized design plans, thereby rendering the houses affordable to customers with the targeted middle-income and upper middle-income range of $80,000 to $125,000.

Existing Human Resources System

Objectives. The Company has attempted to formulate an employee policy that provides detailed job description and fair hiring practices, periodic training programs, employee incentive and benefit programs, and profit sharing. The Company believes the employees are the social capital of a successful company, and through investment in its human resources it will receive a substantial rate of return. Therefore, keeping the employees satisfied with their work and current with new technologies is of prime concern. The task is to employ people who enjoy and are challenged by what they do for the Company and reward them for their achievements that also benefit the Company.

Training. Sales personnel are trained by the Company, and they attend periodic meetings. They are to be updated on sales techniques, competitive products in the area, the availability of financing, construction schedules, and marketing and advertising plans, which management believes results in a sales force with extensive knowledge of the Company's operating policies and housing products.

Superintendents and managers are periodically updated on all the major elements of job-site management that affect project performance. Setting priorities, project planning, people building, maintaining communications at all levels, and building goodwill with contracted help are areas of management that when properly applied will make a difference in a project's economical and timely performance.

Existing Corporate Strategy

Corporate Objective. As a service to our community we will provide an excellent house-product which will meet our homebuyers' increasing requirements of both high quality and affordability. The Company believes a marketable product at a reasonable profit will result from the employment of an efficient decentralized organizational approach, the utilization of current technology, and the application of sound financial principles.

Goals. To achieve this objective, the Company has devised the following operational home-building goals:

1. *Concentration on the development and construction of high-quality, medium-priced, single-family homes for use as primary residences.* The Company has held to the belief that the market for primary residences is more resistant to economic downturns than the market for second or vacation homes. The average selling price of the Company's homes increased in 20XX to $309,407 from $279,073 the previous year. This increase was primarily due to the rising cost of land and building materials. The Company has concentrated its efforts on acquiring land suitable for construction of homes generally in the price range of $250,000 to $415,000, which has represented a broad market segment in the Company's market areas.

2. *Keeping construction in line with anticipated demand.* The Company customarily has acquired unimproved land zoned for residential use that appeared suitable for the construction of 20 to 100 homes in increments of 5 to 20 homes. By developing projects in increments, the Company has been able to control the number of its completed and unsold homes.

3. *Geographic diversification.* The Company has concentrated its home-building activities in Snohomish, Pierce, and King Counties in western Washington. Additionally, the Company now has projects in Kitsap, Thurston, and Mason Counties. The Company's policy of diversifying among different geographic areas has enabled it to reduce the impact of adverse local economic conditions.

4. *Use of local managers for the home-building divisions.* The Company has employed local managers, who initiate land purchases, work with local subcontractors, and generally manage the Company's home-building projects as separate profit centers. The use of local managers has enabled the Company to benefit from the advantages of in-depth local expertise.

5. *Continuing emphasis on the control of overhead and operating expenses.* The Company has traditionally sought to minimize overhead expenses in an effort to control costs and be more flexible in responding to the cyclical nature of its business. Additionally, the Company has attempted to control its land acquisition expenses by generally completing a purchase of land only when it can begin construction within a specified time period.

6. *Use of marketing and sales efforts.* The Company's homes generally have been sold by its own staff of sales personnel through the use of model homes which are maintained at each project site. The Company also has made extensive use of advertisements in local newspapers, illustrated brochures, direct mail, and on-site displays.

Existing Planning Process

The Company believes that maintaining a strong balance sheet is the most important step a homebuilder can take to manage and plan for the future in a cyclical business. From an operating standpoint, the Company controls its risks by value engineering its homes. As interest rates rise, the homebuyers will still be there, but they will need to focus on value. The Company is committed to providing value through its quality design and construction to ensure the homes are both desirable and affordable to its customers (Figures B.2, B.3, and B.4).

Totals for the year ending December 31, 20XX	Homes Delivered	Average Home Selling Price	Total Number of Projects	Number of Projects in Sales Stage	Building Sites Owned or Controlled	Homes Under Construction	Unsold Homes
Snohomish County	17	$325,675	1	1	129	13	16
Pierce County	12	$296,900	1	1	121	11	14
Kitsap County	13	$260,800	1	1	108	8	9
Mason County	11	$280,600	1	1	131	6	8
King County	14	$385,500	1	1	32	4	3
Thurston County	6	$268,900	1	1	94	3	4
Total	**73**	**$309,407**	**6**	**6**	**615**	**45**	**54**
Totals for the year ending December 31, 20XX-1	82	$279,073	8	6	562	42	43
Totals for the year ending December 31, 20XX-2	97	$260,183	9	5	439	51	25
Totals for the year ending December 31, 20XX-3	88	$239,583	8	5	336	43	18
Totals for the year ending December 31, 20XX-4	81	$221,867	7	4	279	36	15

FIGURE B.2 Homebuilding Data

	For the Year Ending December 31,				
	20XX	20XX-1	20XX-2	20XX-3	20XX-4
	(Dollars in thousands)				
Total Revenue	$18,993	$19,646	$21,521	$18,979	$16,192
Cost of Sales	$16,853	$17,250	$19,130	$16,620	$13,980
Gross Profit	$2,140	$2,396	$2,391	$2,359	$2,212
Selling, General and Administrative Expenses	$1,407	$1,394	$1,226	$1,156	$987
Operating Income	$733	$1,002	$1,165	$1,203	$1,225
Other Income (Expense)	$41	$34	$31	$26	$23
Income before Income Taxes	$774	$1,036	$1,196	$1,229	$1,248
Provision for Income Taxes	$232	$311	$359	$369	$375
Net Income	$542	$725	$837	$860	$873

FIGURE B.3 Income Statement

Assets	20XX	20XX-1	20XX-2	20XX-3	20XX-4
			For the Year Ending December 31, (Dollars in thousands)		
Cash and cash equivalents	$3,956	$4,095	$4,156	$5,253	$5,436
Accounts receivable	$1,313	$1,526	$2,349	$1,937	$2,519
Inventory	$20,403	$20,299	$16,580	$14,226	$9,733
Deferred tax assets	$311	$306	$287	$278	$249
Other current assets	$265	$256	$237	$231	$217
Total current assets	$26,248	$26,482	$23,609	$21,925	$18,154
Property, plant, and equipment, net	$127	$121	$115	$112	$105
Other assets	$234	$227	$214	$207	$194
Total assets	$26,609	$26,830	$23,938	$22,244	$18,453
Liabilities					
Short-term debt	$695	$385	$357	$342	$334
Trade accounts payable	$3,169	$3,054	$3,012	$2,958	$2,877
Interest payable	$1,314	$1,282	$1,068	$961	$696
Other payables and accrued liabilities	$1,995	$1,768	$1,645	$1,447	$1,387
Total current liabilities	$7,173	$6,489	$6,082	$5,708	$5,294
Notes payable	$14,598	$14,244	$11,872	$10,683	$7,732
Total liabilities	$21,771	$20,733	$17,954	$16,391	$13,026

FIGURE B.4 Balance Sheet

Appendix C

CASE STUDY
Northwest Constructors Annual Report

Company Description

History. Northwest Constructors, Inc. (the "Corporation"), is an employee-owned corporation that was incorporated in Seattle, Washington, in 1985. The Corporation was organized under the laws of the State of Washington to succeed to the business of Pacific Northwest Builders, which was founded in 1925 by Alfred R. Andersen as a sole proprietorship. The Corporation has engaged in the construction business since 1925. Over the years, the Corporation has constructed a wide range of office, institutional, commercial, industrial, and public buildings, among the most notable being Swedish Hospital Addition, Bellevue Community College, Sea Tac Airport Additions, J. C. Penney retail outlets at local malls, Boeing Company Training Center, and Century Square in Seattle.

Overview. Although the Corporation does work throughout much of the Pacific Northwest, it concentrates its activities principally in the Puget Sound Region. Its executive offices are located at High Rise Plaza in Bellevue. The Corporation presently operates in commercial construction work, engaging in all types of general construction, including medical and educational facilities, public institutions, retailers, industrial operations, commercial operations, and office towers. In addition, the Corporation provides project management and construction management services on a cost-plus-fee basis to a range of infrastructure projects. As a general contractor, the Corporation provides construction services in accordance with the terms and specifications of each contract, including planning and scheduling, marshaling of manpower, procurement of equipment and materials, awarding of subcontracts, and direction and overall management of the project. Generally, purchasing of materials and services for the Corporation's construction operations is done on a project-by-project basis. The Corporation's earnings during 20XX and the previous two years were $3,850,000, $5,278,000, and $9,126,000, respectively.

Employment. Total Corporation employment varies since it depends upon the volume, the type and the scope of operations under way at any given time, as well as upon weather conditions and other factors. On December 31, 20XX, the Corporation employed 309 salaried employees, which included administrative, professional, and executive personnel; 36 hourly paid foremen, and 94 hourly project direct-hire craft employees. On its construction jobs, the Corporation utilizes local labor whenever practical, paying the prevailing wage scale.

Properties. The Corporation occupies 22,500 square feet of usable space in various Corporation-owned buildings in Seattle. Principal facilities owned by the Corporation include a 100,500–square feet office building which the Corporation partially occupies, as well as a 30,250–square feet warehouse and maintenance complex located in Kent, Washington, on 6.9 acres of land. In addition, the Corporation presently owns more than 230 units of heavy and light construction equipment. The Corporation considers that its construction equipment, manufacturing facilities, and administrative properties are well maintained and suitable for its current operations. Maintenance and repair expenses of $219,660 in 20XX have been charged to operations.

Company Organization

The Executive Officers of the Corporation are as follows:

successfully worked on a variety of very complex projects. Combining an undergraduate civil engineering degree from Washington State University with an MBA degree from Seattle University, Mr. Rubin has also directed the Corporation's Marketing and Public Relations divisions.

Executive Officers of the Corporation are elected by the Board of Directors to serve at the pleasure of the Board. The Board of Directors currently consists of six directors who, according to the Corporation's by-laws,

Name and Age	Positions with the Corporation	Officer Since
Bernard W. Andersen (69)	Chairman of the Board of Directors and President (Grandson of Alfred R. Andersen)	1987
Edward P. Abrams (61)	Treasurer and Vice President for Administration	1990
Joseph P. Rubin (51)	Secretary and Vice President for Marketing and Public Relations	1994

Brief Biography of Executive Officers
Bernard W. Andersen. Mr. Andersen, grandson of the founder, grew up in the construction business. After obtaining a bachelor of science in building construction at the University of Washington, he worked at nearly every position in over 35 years of construction experience. Mr. Andersen has been directly responsible for projects of all sizes and in a wide variety of types. Today, he oversees all Corporation operations.

Edward P. Abrams. Mr. Abrams has more than 30 years of experience in construction management and preconstruction services. Mr. Abrams' wide construction background combined with an accounting degree from Stanford University has made it possible to successfully manage the Corporation's administrative functions. Recently, Mr. Abrams has instituted a company-wide information system, which has increased the communication efficiency of the Corporation.

Joseph P. Rubin. In his 20 years with the Corporation, Mr. Rubin's duties have included assignments as a project manager and a construction manager and he has

are divided into three classes. The provision is designed to ensure that approximately one-third of the Board is eligible for reelection in any given year. The Board of Directors has established standing committees, including an Audit Committee, a Compensation Committee, and a Nominating and Organization Committee. All committees are responsible to the full Board of Directors.

Each project is managed by a project manager whose assistants, depending upon job complexity, include a field engineer and a job superintendent to supervise actual construction. Often assigned to more than one project, the manager is responsible to the Vice President for Construction Operations, who reports to the President. Also reporting directly to the President are the Vice President for Administration and the Vice President for Marketing and Public Relations, each responsible for the operations of their respective corporate division.

The construction manager usually directs the work on projects involving multiple prime contractors and vendors. The construction manager's responsibilities are defined by the construction management contract with the owner. Typically, the construction manager supervises a project's quality control, checks and updates the time schedule, reviews the change orders, and keeps the owner and architect informed if a potential cost overrun may occur. He also runs the job meetings, keeping and distributing records to all involved parties. The construction manager is responsible to the Vice President for Construction Management, who in turn reports to the President.

Corporate Organization Chart

Both as a general contractor and a construction manager, the Corporation engages its personnel in supervising and coordinating work on the projects (Figure C.1). Although subcontractor services are employed to perform much of the physical development and construction work on the projects, the Corporation is often also involved in the construction of the concrete or steel superstructure of the buildings. Because of its long-standing market presence, the Corporation has developed strong relationships with a number of prime subcontractors who provide reliable quality service. Contractual agreements with subcontractors are typically accomplished with a fixed price after competitive bidding.

In the case of projects where the Corporation acts as general contractor rather than

FIGURE C.1 Organization Chart

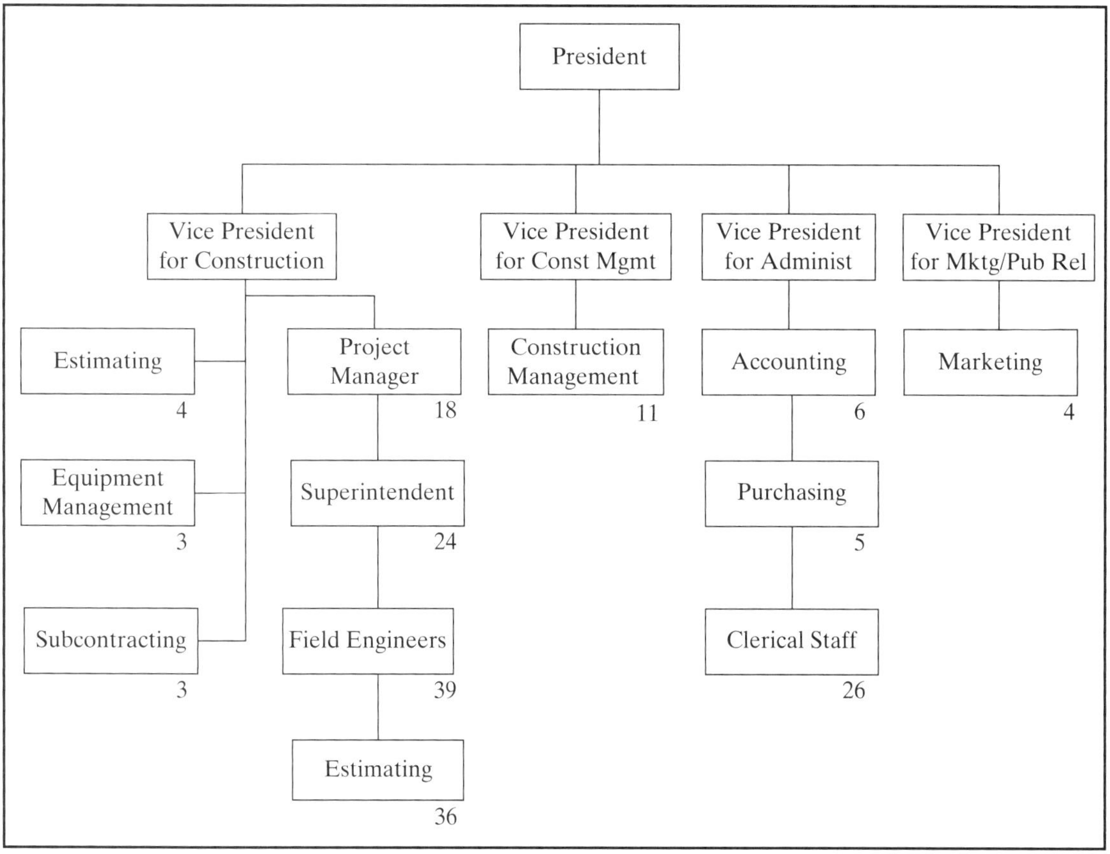

construction manager, it is usually required to furnish payment and performance bonds. The Corporation believes its bonding capacity is adequate for both present and future project requirements. The aggregate bonding capacity of the Corporation on December 31, 20XX, was $225,000,000.

Company Services

Preconstruction Services. The Corporation firmly believes that properly accomplishing a project requires careful preconstruction planning. A personalized approach involving the team in the design phase will match the Corporation's management resources with the unique characteristics of the project, helping to assure success. Utilizing cutting-edge computer software, the Corporation's early design phase cost models which employ abundant historical data will help determine a project's feasibility. These models present an early basis for creating detailed estimates as well as comprehensive project budgets in later phases of project development. With an average of over 15 years of experience per estimating staff member, competitive and reliable estimates have been possible.

In addition, specifications and details for each project are reviewed by the project management to determine constructability. Specifications are examined to assure that the materials fit their intended use, they are technically compatible with contiguous materials, and they interface correctly with all building systems. As a result, potential changes that will facilitate building fabrication as well as reduce long-term maintenance costs will be offered to the architect and the owner.

Value Engineering. Using historical data generated over decades, each division of work on a project is analyzed to determine the most cost-effective fabrication techniques and materials. Driven by the goal of optimizing project success, systems analysis is employed to recommend alternative systems that support the overall design intent and provide the best value. Correctly employed, value engineering serves the dual purpose of bringing a project within budget while maintaining quality, and reducing the Corporation's operating risks.

Scheduling. On all projects, the project managers and superintendents combine their practical knowledge and scheduling skills to prepare realistic and suitable work sequences. A full-time scheduling team becomes involved with regular project reviews and updates. In these meetings, the project's unique characteristics are defined and considered in a procedure which includes submittal tracking, resource leveling, cash-flow analysis, and other appropriate functions. Special needs unique to the project are also considered; such as, the possibility of work around occupied space, special events, ordering items that require long lead times, and installation of owner-purchased items, to name a few.

Project Management. The Corporation is committed to a genuine team approach in its projects. To facilitate project construction, on-site crews are notified of direct communication links with local management as well as home office staff. In addition, each jobsite superintendent is expected to meet weekly with home office management to address scheduling concerns. Also routine are the meetings involving the Corporation's local project management team with the owner and the architect.

The local project management staff handles day-to-day administrative functions including the following:

- Submittal and shop drawing review
- Schedule updating
- Quality control
- Crew and subcontractor management
- Jobsite safety
- Field questions processing
- Change order processing
- Project closeout

Construction Management. Before construction occurs, the construction manager (CM) is actively involved in reviewing the design for constructability and cost effectiveness. Value engineering efforts will result in substantial savings to the owner in terms of both time to construct and total construction costs. During construction, the CM plays a major role in project scheduling, payment requisition review, and change order analysis. The last responsibility is particularly important, because the CM will look more objectively at the design and potential ambiguities in it than would the design professional on the project.

Projects that are managed by the Corporation's CM usually result in a more cost-effective product to the owner than the traditional method of project delivery. However, the fundamental relationships among the parties remain the same because the design and construction functions are performed by separate entities, even though there is more preconstruction involvement with a CM than under the traditional method of contracting.

Safety. The Corporation recognizes that an excellent safety record results in reduced construction costs. The Corporation's low incident accident rate entitles it to lower-than-average workers' compensation insurance rates, reducing overall labor costs for each project. To maintain this positive safety record, a proactive approach has been taken which enhances the protection of employees, subcontractors, clients, and the public. For example, a complete safety program has been developed which requires every project team member to participate. Each project also has a safety coordinator, who conducts weekly jobsite safety inspections and meetings. Every employee is involved in safety awareness programs and is encouraged to contribute ideas to improve the Corporation's record. Finally, home office management periodically reviews all safety policies and procedures to keep safety in the forefront of every project's operations. Safety awards are given to superintendents who complete projects without lost-time accidents.

Post-Construction Services. The Corporation's services are designed to make its clients comfortable with every aspect of their new facility. At the completion of the job, the project staff guides the owner through the new building step by step to familiarize them with each system's operational requirements and ascertain that all systems are functioning properly. Ultimately, the objective of each project is to produce a trouble-free building which will perform without interference to the owner. When warranty claims arise, the Corporation is committed to responding quickly, backing its product.

Current Company Trends

Risk Management. In order to balance risk with reward, the Corporation enters into three basic types of contracts:

1. Fixed-price contracts providing for a single price for the total amount of work to be performed;
2. Cost-plus-fee contracts with guaranteed maximum costs, under which risk is reduced and anticipated income is accordingly lower;
3. Cost-plus-fee contracts, under which risk is minimal and anticipated income is earned solely from the fee received for services provided.

The table on p. 214 shows the percentage of work done under each type of contract during the last five years.

Regardless of the type of contract, the Corporation has always been subject to unusual risks, including unforeseen conditions encountered during construction, the impact of inflation upon costs and upon financing requirements of clients, and changes in political and legal circumstances, especially since contracts for major projects

Type of Contract	Current Year	Current Year-1	Current Year-2	Current Year-3	Current Year-4
Fixed-price	23%	28%	35%	43%	47%
Cost-plus-fee w/guaranteed max.	45%	39%	33%	26%	23%
Cost-plus-fee	32%	33%	32%	31%	30%

are performed over an extended period of time. Although the Corporation constantly seeks to minimize and spread the risks over a large number of contracts, a combination of unusual circumstances could result in a loss on a particular contract.

The Corporation is also responsible for any failure to perform on the part of a subcontractor. In order to minimize the potential for losses caused by such defaults, the Corporation normally prequalifies subcontractors and requires performance and payment bonds.

Seasonality. The Corporation's business, generally, has been seasonal, with the highest revenues usually occurring in the second and fourth quarters of the calendar year. This seasonal nature results from weather factors as well as clients' desires to build or remodel retail outlets so that work will be completed by spring or fall.

Joint Ventures. The Corporation often participates, usually as manager, in construction joint ventures with others to share risks and combine financial, technical, and other resources. Each of the joint venture participants is usually committed to supply a predetermined amount of capital, and to share in a predetermined percentage of income or loss of the joint venture. Joint ventures normally have a short life span, since they are designed and created for the sole purpose of bidding on, or negotiating for, and completing one specific project. These single-purpose joint ventures last only as long as the construction project is undertaken, which can be less than one year. Construction joint ventures undertaken to complete a specific project are liquidated when the project is completed.

Regulation. The Corporation is subject to the authority of various state and local regulatory agencies concerned with the licensing of contractors, but it has experienced no material difficulty in complying with such requirements. The Company is also subject to local zoning regulations and building codes in performing its construction activities. Compliance with regulatory provisions which have been enacted for regulating the discharge of materials in the environment have not had a material effect upon the capital expenditures, earnings, and competitive position of the Corporation.

New Business and Backlog. Backlog consists of uncompleted portions of construction contracts, including the Corporation's share of construction joint venture contracts. Nearly 68 percent of the 20XX year-end backlog is expected to be completed next year. Since backlog reflects only business which is considered to be firm and is an indication of expected future revenues, there can be no assurance that cancellations or scope adjustments will not occur or when revenue from such backlog will be realized. The Corporation's estimated backlog for uncompleted construction projects was $142.8 million at December 31, 20XX, as compared to $173.5 million and $211.0 million at the end of the previous two years. The Corporation is actively seeking to increase its backlog of business, especially in the government and institutional sectors.

Description of Major Clients

The Corporation is a full-service contractor, offering preconstruction, general contracting, and construction management services for medical, educational, hospitality, retail, industrial, and commercial clients. Over the years the Corporation has completed over 1,200 projects with a value in excess of $1,400 million. For the past three years average yearly revenues have exceeded $206 million. Although the Corporation's primary market is the Puget Sound Basin, the region that lies between Olympia and Bellingham, projects have also been constructed in Alaska, Oregon, Montana, Idaho, and California. Though known for larger projects, the Corporation's experience has included jobs of almost any size and with a wide range of recognizable clients:

Hospitals and Medical Facilities

- Ballard Community Hospital
- Children's Hospital and Medical Center
- Group Health Cooperative of Puget Sound
- Overlake Hospital Medical Center
- University of Washington Teaching Hospital
- Valley Medical Center

Educational and Public Institutions

- Battelle Seattle Research Center
- Bellevue Community College
- Bremerton School District
- City of Seattle
- Edmonds Community College
- King County
- Seattle Housing Authority
- Seattle School District
- Seattle University
- U.S. Navy
- University of Alaska
- University of Washington
- State of Washington
- Washington State University

Hotels and Retailers

- Albertson's
- Allied Stores
- Bon Marche
- Frederick and Nelson
- J. C. Penney
- Safeway
- Sears

Industrial Operations

- Boeing Company
- Crown Zellerbach
- Immunex
- Pacific Coca-Cola Bottling
- Port of Seattle
- Sealaska
- Seattle Times
- United Airlines

Commercial Buildings and Office Towers

- Bank of California
- Century Square
- First Interstate Bank
- IBM
- Pacific Northwest Bell
- Washington Athletic Club
- Washington Mutual

Construction Management

- Swedish Hospital Medical Center
- Harborview Medical Center
- Belltown Apartments
- Federal Aviation Administration
- Fifth Avenue Theater Renovation
- Puyallup High School
- Sea Tac Airport
- Seattle Sheraton
- Seattle Supersonics Training Facility
- Unico Properties

Description of Competition

The Corporation is engaged in a highly competitive business, especially that portion which relates to contracts for construction obtained by open bidding. The Corporation competes with other general and specialty contractors, including a number of small- to moderate-size local contractors.

Existing Corporate Strategy

Corporate Objective. Now in its seventh decade, the Corporation and its predecessor company have seen the construction of hundreds of structures throughout the Northwest. Today it is one of the major general contracting companies in the Pacific Northwest and yet it treats each project as unique with a specific set of challenges—not as just another job. The Corporation's success is owed to its founder, Alfred R. Andersen, who took a personalized service approach in his projects by staying in close contact with his clients during and after the job, and whose phrase *"Do it right the first time"* encapsulated his dedication to construction excellence.

Goals. The Corporation's strategy of being the best, by exceeding client expectations while adhering to the highest construction standards, will be accomplished by fulfilling the following goals:

1. The Corporation will provide construction services of the highest possible quality to its clients.
2. The Corporation's services will be organized and delivered to be fully responsive to the needs of clients and prospects.
3. Particular emphasis will be placed on cost-effectiveness, so that clients always receive full value for their facility investments.
4. The Corporation will strive to develop unique capabilities and qualifications to give it a significant competitive advantage in those markets in which it competes.
5. The Corporation will strive to increase productivity by stimulating and encouraging innovation, as well as through the application of advanced technologies.
6. The Corporation will encourage excellent and enthusiastic performance by employees through appropriate polices governing compensation, incentives, promotion, benefits, and challenging assignments.
7. In order to apply the Corporation's technical and human resources with optimum efficiency, managers will be carefully selected, appropriately trained, and fully motivated.
8. In a world of fast-changing markets and rapidly emerging technology, the Corporation will remain competitive by vigorously refining and expanding its technical and professional capabilities.
9. The nature of this business requires a strong financial base. At all times, the Corporation will retain sufficient capital resources to meet its commitments. For future growth, it will make substantial investments to develop new services and apply new technologies.
10. The Corporation is committed to uncompromising values in its corporate conduct. Integrity, in the broadest sense, will guide the Corporation's actions in all relationships, whether with employees, clients, suppliers, or the government. The Corporation will always comply fully with all laws and regulations.

Existing Human Resources System

Human Resources Objective. The Corporation is striving to expand its client base in the Pacific Northwest. To achieve this, the management is working to raise the quality of its professional, technical, administrative, and support staffs. This commitment to increasing personal excellence in the performance of its workforce is the cornerstone of the Corporation's human resources system.

Recent efforts to enhance productivity "inside the construction fence" are an outcome of the Corporation's commitment to its personnel. "*Focus groups*," which consist of key field and home office staff, tackle topics such as procedures for enhancing project planning, improving project cost history, arranging effective purchasing agreements, and more effectual subcontractor prequalification procedures. As a result of addressing each of these and other issues on a regular basis, additional productivity has been realized, including economies due to improved computer hardware and updated software. To further enhance personnel effectiveness while providing more value to the client, this program will continue.

An additional program, the "*Construction Idea of the Month*," is causing additional benefits to occur while enhancing the workplace for the personnel. Every month a job field crew is recognized for its ideas which contribute toward enhanced productivity and lower project costs. As an acknowledgment of the contribution, the winning team is treated to lunch by the corporate officers who are also required to visit the team's jobsite during the event.

Training. One of the roles of the Vice President for Construction Operations is to assure that excellent training programs and materials are developed for use on projects, especially in the category of safety management. In addition he must make sure that jobsite managers follow through and institute the programs to train new workers and foremen in procedures unique to the Corporation. Superintendents and managers must also be updated on such topics as promoting communications at all job levels, building goodwill and a sense of team effort, enhancing project planning and priority-setting skills, and building up personnel. Periodic meetings are held which touch on those elements of jobsite management, which most directly impact project performance. A company-wide performance evaluation program has been implemented to identify and address personnel issues with corrective training.

Employee Benefits. The Corporation provides retirement benefits for nearly all employees, either through a defined benefit plan, the defined contribution Employee Stock Ownership Plan (ESOP), or a combination of both types of plans. The Corporation bases pension contributions on funding standards established by the Employee Retirement Income Security Act of 1974. The majority of the Corporation's salaried and hourly employees are covered by both the defined benefit plans and the ESOP. All stock owned by an employee must be sold back to the Corporation upon the employee's separation whether by resignation or by retirement.

In addition, the Corporation sponsors the Andersen Employees Savings Trust, a savings plan under section 401(k) of the IRS. All salaried and nonunion hourly employees are eligible to participate in this plan. Under its terms the Corporation will match 25 percent of the first 6 percent of employees' contributions to the plan.

The Corporation also provides health care and life insurance benefits for retired employees and their dependents. Substantially all of the Corporation's employees become eligible for these benefits if they reach retirement age while still working for the Corporation and have at least ten years of service.

Project managers and superintendents are updated on the major principles of on-site management that impact project and safety performance. Management principles having to do with team building, developing and maintaining communications, setting project priorities, and developing strong ties with contracted help will, when applied, make a significant difference in a project's successful performance.

Current Marketing Activities

The Corporation's employees who have daily contact with its current and potential customers are best able to alert the management to client needs and opportunities. While construction practice operates largely in a team environment, emphasis is nonetheless focused on every person's individual responsibility in achieving the marketing goals of the Corporation. Personnel are required to attend periodic meetings to be updated on techniques for creating strong client relations.

This past year was one of many challenges for the Corporation. This year will be one in which the Corporation will establish its expansion partly by pursuing opportunities in new areas of construction management. The Corporation's construction management division is now expanding its expertise into hospital and medical research facilities with projects like the Gaittes Research Laboratories at Memorial Cancer Center and the Seattle Psychiatric Institute Addition. Management has been exploring this field and evaluating the competition with the goal of expanding the Corporation's construction management portfolio into Alaska and California.

Current Planning Process

To ensure controlled growth, the Corporation has set formal controls in place. These include strict budgeting, control of cash, and regular reporting. Since budget control is an essential tool for managing a successful project, the Corporation's overall budgeting process is intensive. A centralized cash management system means it is possible to know daily how much cash is being used or provided at each operation. The reporting systems cover all aspects of the projects and are reviewed by the management in regular visits to the jobsites (Figures C.2, C.3, C.4, C.5, C.6, and C.7).

FIGURE C.2 Status of Projects in Progress

	Total Revenues on December 31, 20XX (in thousands)			
	Contract Price	Actual to Date	Estimated to Complete	Total
Completed Projects				
Swedish Hospital Addition	$30,543	$31,652		$31,652
Bremerton High School	$6,532	$6,744		$6,744
Edmonds Community College Additions	$11,244	$11,750		$11,750
Albertson's Grocery Store	$2,219	$2,211		$2,211
Bellevue Athletic Club	$14,750	$14,894		$14,894
Bellingham Sheraton Hotel	$18,659	$18,645		$18,645
Highline Performing Arts Center	$9,327	$10,348		$10,348
Allenmore Hospital	$42,890	$42,976		$42,976
Frank Baker Surgey Center	$33,472	$34,277		$34,277
Total Completed Projects	**$169,636**	**$173,497**		**$173,497**
Projects in Progress				
Pierce Transit Administrative Office	$24,561	$6,866	$17,788	$24,654
Target Retail Store in Puyallup	$11,450	$8,452	$3,000	$11,452
Pacific Northwest Bell Building	$21,875	$7,330	$14,545	$21,875
Sparks Stadium	$3,890	$2,998	$856	$3,854
Western Washington Fair Addition	$16,942	$9,006	$7,988	$16,994
Goodman Middle School	$7,967	$5,703	$2,250	$7,953
Total Projects in Progress	**$86,685**	**$40,355**	**$46,427**	**$86,782**
Combined Totals	**$256,321**	**$213,852**	**$46,427**	**$260,279**

(Dollars in thousands, except per share amounts)	20XX	20XX-1	20XX-2	20XX-3	20XX-4
Consolidated Operating Results					
Revenues	$188,852	$205,306	$224,700	$211,375	$195,550
Earnings from continuing operations, before taxes	$4,853	$6,603	$11,333	$10,882	$9,968
Net earnings	$3,850	$5,278	$9,126	$8,996	$8,159
Earnings per share	$0.67	$0.88	$1.50	$1.41	$1.30
Return on average stockholders' equity	14.90%	18.76%	43.09%	35.63%	37.70%
Cash dividends per common share	$0.16	$0.21	$0.37	$0.32	$0.24
Consolidated Financial Position					
Current assets	$49,571	$51,334	$49,853	$47,568	$42,876
Current liabilities	$29,546	$29,046	$28,620	$28,110	$27,333
Working capital	$20,025	$22,288	$21,233	$19,458	$15,543
Property, plant and equipment, net	$15,764	$14,862	$14,358	$13,320	$12,420
Total assets	$66,321	$66,897	$64,798	$61,481	$55,814
Capitalization					
Long-term debt	$10,735	$8,497	$6,276	$5,124	$4,063
Shareholders' equity	$25,835	$28,134	$26,773	$25,249	$21,643
Total capitalization	$36,570	$36,631	$33,049	$30,373	$25,706
Percent of total capitalization					
Long-term debt	29.35%	23.20%	18.99%	16.87%	15.81%
Shareholders' equity	70.65%	76.80%	81.01%	83.13%	84.19%
Other Data					
New awards	$96,443	$128,832	$165,922	$161,655	$159,210
Backlog at year end	$142,870	$173,452	$210,972	$201,551	$196,723
Capital expenditures	$10,149	$12,083	$14,930	$14,560	$12,223
Cash provided by operating activities	$12,291	$15,015	$17,844	$17,926	$14,278
Salaried employees	309	332	396	431	412

FIGURE C.3 Selected Financial Data

(Dollars in thousands)	20XX	20XX-1	20XX-2	20XX-3	20XX-4
Cash Flows from Operating Activities					
Net earnings	$6,505	$7,637	$8,988	$8,455	$7,822
Total cash provided by operating activities	$12,291	$15,015	$17,844	$17,926	$14,278
Cash Flows from Investing Activities					
Capital expenditures	$10,149	$12,083	$14,930	$14,560	$12,223
Investments	$65	$130	$156	$181	$280
Proceeds from sale of property, plant and equipment	$788	$923	$556	$856	$512
Other, net	$386	$463	$439	$387	$378
Cash utilized by investing activities	$11,388	$13,599	$16,081	$15,984	$13,393
Cash Flows from Financing Activities					
Payments on long-term debt	$1,288	$1,020	$753	$615	$488
Short-term borrowing	$36	$54	$59	$64	$45
Cash dividends paid	$304	$322	$353	$312	$287
Stock options exercised	$92	$107	$122	$109	$97
Other, net	$18	$22	$27	$43	$37
Cash utilized by financing activities	$1,738	$1,525	$1,314	$1,143	$954
Cash and cash equivalents at beginning of year	$2,392	$2,501	$2,052	$1,253	$1,322
Cash and cash equivalents at end of year	$1,557	$2,392	$2,501	$2,052	$1,253

FIGURE C.4 Consolidated Statement of Cash Flows

(Dollars in thousands)	20XX	20XX-1	20XX-2	20XX-3	20XX-4
Total Revenues	$188,852	$205,306	$224,700	$211,375	$195,550
Cost of Revenues	$178,391	$192,465	$206,750	$194,355	$179,614
Other (Income) and Expense					
Corporate administrative and general expense	$5,084	$5,836	$6,360	$5,945	$5,812
Interest expense	$650	$520	$370	$300	$250
Interest income	($126)	($118)	($113)	($107)	($94)
Total Expenses	$183,999	$198,703	$213,367	$200,493	$185,582
Earnings from Continuing Operations before Taxes	$4,853	$6,603	$11,333	$10,882	$9,968
Income Tax Expense	$1,003	$1,325	$2,207	$1,886	$1,809
Earnings from Continuing Operations	$3,850	$5,278	$9,126	$8,996	$8,159
Loss from Discontinued Operations, Net				($376)	($158)
Net Earnings	$3,850	$5,278	$9,126	$8,620	$8,001
Earnings per Share	$0.67	$0.88	$1.50	$1.41	$1.30

FIGURE C.5 Consolidated Statement of Earnings

(Dollars in thousands)	20XX	20XX-1	20XX-2	20XX-3	20XX-4
Assets					
Current Assets					
Cash and cash equivalents	$1,557	$2,392	$2,501	$2,052	$1,253
Marketable securities	$1,172	$1,221	$1,522	$1,233	$1,510
Acounts and notes receivable	$19,527	$21,428	$20,729	$19,974	$18,994
Contract work in progress	$25,819	$24,752	$23,827	$22,497	$19,274
Inventories	$567	$453	$432	$365	$389
Net assets of discontinued operations				$740	$351
Deferred taxes	$588	$665	$675	$534	$477
Other current assets	$341	$423	$167	$173	$628
Total current assets	$49,571	$51,334	$49,853	$47,568	$42,876
Property, Plant and Equipment					
Land	$598	$564	$534	$521	$499
Buildings and improvements	$3,134	$3,042	$2,877	$2,695	$2,532
Machinery and equipment	$17,276	$16,385	$15,924	$14,854	$13,958
Subtotal	$21,008	$19,991	$19,335	$18,070	$16,989
Less accumulated depreciation, depletion,and amortization	$5,244	$5,129	$4,977	$4,750	$4,569
Net property, plant and equipment	$15,764	$14,862	$14,358	$13,320	$12,420
Other Assets					
Investments	$715	$523	$422	$450	$395
Other	$271	$178	$165	$143	$123
Total other assets	$986	$701	$587	$593	$518
Total Assets	**$66,321**	**$66,897**	**$64,798**	**$61,481**	**$55,814**

FIGURE C.6 Consolidated Balance Sheet

(Dollars in thousands)	20XX	20XX-1	20XX-2	20XX-3	20XX-4
Liabilities and Shareholders' Equity					
Current Liabilities					
Accounts payable	$13,002	$12,890	$12,653	$12,475	$12,122
Advance billings on contracts	$12,064	$11,986	$11,887	$11,730	$11,529
Accrued salaries, wages, and benefit plan liabilities	$1,995	$1,942	$1,911	$1,877	$1,856
Other accrued liabilities	$2,105	$1,907	$1,879	$1,750	$1,561
Current portion of long-term debt	$380	$321	$290	$278	$265
Total current liabilities	$29,546	$29,046	$28,620	$28,110	$27,333
Long-Term Debt Due After One Year					
Long-Term Debt Due After One Year	$5,355	$4,176	$3,986	$2,834	$1,798
Noncurrent Liabilities					
Deferred taxes	$588	$665	$675	$534	$477
Other	$4,997	$4,876	$4,744	$4,754	$4,563
Total noncurrent liabilities	$5,585	$5,541	$5,419	$5,288	$5,040
Total Liabilities	$40,486	$38,763	$38,025	$36,232	$34,171
Shareholders' Equity					
Capital stock	$4,989	$4,785	$4,645	$4,329	$4,255
Retained earnings	$20,846	$23,349	$22,128	$20,920	$17,388
Total shareholders' equity	$25,835	$28,134	$26,773	$25,249	$21,643
Total Liabilities and Shareholders' Equity	**$66,321**	**$66,897**	**$64,798**	**$61,481**	**$55,814**

FIGURE C.7 Consolidated Balance Sheet

Appendix D

Employment Manual for Western States Construction Company

Introduction

On behalf of your coworkers, we welcome you to Western States Construction Company. Our primary goal is to provide outstanding service to our customers. We have a reputation for high-quality services; that reputation was built and is maintained by the efforts of each one of our employees. We expect all employees to uphold and add to our excellent reputation. We want people who take pride in their work and themselves. This pride of workmanship enhances the company's reputation.

Our employees are our most valuable assets. While our vision and passion drive us toward our future, relationship-building with our customers, community, coworkers, supervisors, and vendors is the core of our business. Through your efforts as part of our team, we provide a quality service at a fair price to our customers. Pride in your work and your company is a key ingredient in our future success.

General Employment Policies

Employment Relationship
You joined Western States Construction Company voluntarily and are free to resign at any time. Similarly, the Company is free to end our employment relationship when we believe it is in the Company's best interest. Accordingly, *all employees are employees at will and may be terminated at any time with or without notice and with or without cause.* Neither the employee nor the Company has entered into any express or implied contract of employment. No Company representative has the authority to change this At-Will relationship, except in writing signed by the employee and the President of the Company.

The 1986 Federal Immigration laws require all employers to verify the eligibility of employees not later than their third day of employment. You are required to complete Form I-9, which will be countersigned by a Human Resources staff member who will check your documentation. The acceptable forms of documentation are listed on the form.

This section is intended to define the typical employment relationships present at the Company so you can identify your employment status and benefit eligibility.

Exempt and Nonexempt: Each employee is designated as either *exempt* or *nonexempt* from federal and state wage and hour laws. *Nonexempt* employees are entitled to overtime pay under the specific provisions of federal and state laws. An employees' *exempt* or *nonexempt* classification may be changed only upon written notification by the Company management.

Probationary and Regular: Newly hired employees begin a *probationary* period of six months to confirm their decision to join the Company team. During this period you have the opportunity to better understand your suitability for the job. In the same way your supervisor or manager can assess the hiring decision to confirm your abilities and potential for the job. This period may be ended

APPENDIX D   Employment Manual for Western States Construction Company

> **Please Remember that our Future Depends on You**
>
> This manual is intended to provide you with a general understanding of the personnel policies at Western States Construction Company. You should familiarize yourself with the contents of this manual, for it will answer many common questions concerning employment with the Western States Construction Company. Individuals who are members of one of the bargaining units should refer to their union contract for negotiated conditions of employment. If any of the policies are in conflict with the Labor Agreements signed by Western States Construction Company, the Labor Agreement takes precedence for employees covered by that agreement. *This manual is not an employment contract and is not intended to create contractual obligations of any kind.*
>
> The Company cannot anticipate every situation or answer every question about your employment. In order to retain necessary flexibility in the administration of policies and procedures, Western States Construction Company reserves the rights to change, revise, or eliminate any of the policies and/or benefits described in this manual, except for its policy of employment-at-will. The only recognized deviations from the stated policies are those authorized and signed by the President of Western States Construction Company.
>
> *Nothing contained in this manual is intended to be a part of your employment relationship. Rather, the contents of this handbook are simply general statements of company policy and are not promises of specific treatment.*
>
> You will be asked to read and sign an acknowledgment form as a way of confirming your commitment to review the manual in its entirety. Once you have read this manual, you are encouraged to raise any questions that you may have with your supervisor.

before six months in termination of employment or may be extended beyond six months if considered necessary. Once you have successfully completed your *probationary* period you become a *regular* employee eligible for benefits described later in this manual. The transition to *regular* employment status does not change the "*at-will*" employment relationship.

Full-Time and Part-Time: If you are scheduled to work at least 30 hours a week, you are considered a *full-time* employee and are eligible for the Company benefit package, subject to the terms, conditions, and limitations of each benefit program. If you are scheduled for less than 30 hours a week, you are considered a *part-time* employee and will receive benefits mandated by law (e.g. Social Security, Worker's Compensation). *Part-time* employees are not eligible for Company benefits unless specified in this manual.

Temporary: If you have been hired as an interim replacement, to temporarily supplement needs, or to assist in the completion of a specific project, you will be classified as a *temporary* employee. Employment assignments in this category are of limited duration. Like *part-time* employees you are eligible to receive all legally mandated benefits, but not eligible for any of the Company's other benefit programs.

Casual, On-Call: Employees who have established an employment relationship with the Company but who are assigned to work less than 20 hours a week and/or on an intermittent and/or unpredictable basis are considered *casual, on-call*. Benefits for this category are the same as *temporary* employees.

Equal Employment Opportunity Policy

Western States Construction Company maintains a policy of nondiscrimination with all employees and applicants for employment. All aspects of employment with the Company will be governed on the basis of merit, competence, and qualifications and will not be influenced in any manner by race, color, creed, religion, ancestry, sex, age, marital status, national origin, disability, veteran status, or any other legally protected characteristic or status.

All decisions made with respect to recruiting, hiring, and promotion for all job classifications will be based solely on individual qualifications related to the requirements of the position. Likewise, all other personnel matters such as compensation, benefits, transfers, reduction-in-force, training, education, and social/recreation programs will be administered free from any illegal discriminatory practices.

Western States Construction Company is also committed to prohibiting discrimination against qualified individuals with disabilities in job application procedures, hiring, discharge, promotion, compensation, job training, and other terms, conditions, and privileges of employment. The Company will attempt to provide reasonable accommodation to qualified disabled applicants and employees so they can perform the essential functions of a job.

Antiharassment/Anti discrimmination/ Anti-Retaliation Policy

Western States Construction Company is committed to providing a businesslike and professional work environment that is free of harassment, discrimination, and retaliation. The Company policy prohibits sexual harassment and harassment, discrimination, or retaliation based on race, religion, creed, color, national origin, sex, gender, disability, ancestry, age, marital status, veteran status, or any other basis protected by federal, state, or local law or ordinance or regulation. All such harassment is a serious violation of Company policy and may be unlawful. Our policy applies to all persons involved in the operation of the Company and prohibits harassment by any employee of the Company, including supervisors and coworkers, as well as by any person doing business with or for Western States Construction Company. Conduct of Company employees while performing their duties outside the workplace shall also be governed by this policy.

Prohibited harassment, discrimination, and retaliation includes, but is not limited to, the following behaviors:

- Verbal conduct such as epithets, derogatory jokes or comments, slurs or unwanted sexual advances;
- Visual conduct such as derogatory and/or sexually oriented posters, photography, cartoons, drawings, or gestures;
- Physical conduct such as assault, unwanted touching, blocking normal movement, or interfering with work because of sex, race, or any other protected basis;
- Threats and demands to submit to sexual requests as a condition of continued employment, or to avoid some other loss, and offers of employment benefits in return for sexual favors; and
- Retaliation for having reported, threatened to report, or participated in an investigation of harassment or discrimination.

Employees who believe that they have been harassed, discriminated, or retaliated against in violation of this policy must report their concern to their supervisors as soon as possible after the incident. If an employee does not feel comfortable directing this report to his or her supervisor, the report may be directed to the Company President. The complaint should include details of the incident or incidents, names of the individuals involved, and names of any witnesses. Supervisors will refer all harassment, discrimination, or retaliation complaints to the Company President. The Company will immediately undertake an investigation of the harassment, discrimination, or retaliation allegations.

If the Company determines that prohibited harassment, discrimination, or retaliation in violation of this policy has occurred, effective remedial action will be taken in accordance with the circumstances involved. Any employee determined by Western States Construction Company to have engaged in inappropriate conduct in violation of this policy will be subject to appropriate disciplinary action, up to and including termination. The complaining employee will be advised of the results of the investigation and the Company will take appropriate remedial action. The Company will not retaliate against employees for filing a complaint or participating in an investigation, and will not tolerate retaliation by management, employees, or coworkers.

All employees must immediately report any incidents of harassment, discrimination, or retaliation forbidden by this policy so that such concerns can be quickly and fairly resolved. Any report of harassment, discrimination, or retaliation must be made in good faith.

Drug-Free Workplace Policy

Western States Construction Company has a vital interest in maintaining safe, healthful, and efficient working conditions for its employees. Being under any influence of an illegal drug or alcohol on the job may pose serious safety and health risks not only to the user, but to all those who work with the user, as well as our customers. Western States Construction Company also recognizes that its own health and future are dependent upon the physical and psychological health of its employees. Accordingly, the Company has established the following guidelines with regard to use, possession, or sale of alcohol or drugs:

- All applicants and employees will be subject to testing under this policy.
- Western States Construction Company will maintain preemployment screening practices designed to prevent hiring individuals who use illegal drugs or individuals whose use of legal drugs or alcohol demonstrates that the applicant is unable to perform the essential functions of the job with or without reasonable accommodation or poses a significant threat to the health and safety of the applicant or others.
- The manufacture, possession, use, distribution, sale, purchase, transfer, or being under any influence of alcohol or illegal drugs is strictly prohibited while on Company premises or while performing Company business.
- Employees will not be permitted to work while under any influence of illegal drugs or alcohol. Individuals who appear to be unfit for duty may be subject to a medical evaluation, which may include drug or alcohol screening. Refusal to comply with a fitness-for-duty evaluation may result in disciplinary action up to and including dismissal.
- Employees will be tested following an on-the-job injury requiring treatment from a physician, or, following a serious or potentially serious accident or incident, including near misses, in which safety precautions were violated, unsafe instructions or orders were given, vehicles/equipment/property was damaged, or unusually careless acts were performed. Other individuals involved in the incident may also be subject to the drug-testing requirements of this policy.
- Off-the-job use of illegal drugs which could adversely affect an employee's performance or which could jeopardize the safety of other employees, the public, or company facilities, or where such usage could jeopardize the security of Company finances or business records, or where such usage adversely affects customers' or the public's trust in the ability of the Company to carry out its responsibilities will not be tolerated. Employees who are convicted of a crime involving illegal drug activity will be considered in violation of this policy.
- Employees undergoing prescribed medical treatment with a controlled substance that may render them unable to perform the essential functions of the job with or without

reasonable accommodation or that poses a significant threat to the health and safety of the employee or others are required to report this to their manager.
- Employees have the right to request and receive the results of any substance test. All employee information relating to testing will be protected by the Company as confidential unless otherwise required by law or authorized in writing by the employee.

Any employee found in violation of this policy, or who refuses to submit to a drug or alcohol screening, will be removed from Company property or the jobsite and be subject to disciplinary action, up to and including termination of employment.

Western States Construction Company recognizes that alcoholism/drug abuse may be a form of illness that is treatable in nature. The Company will not discriminate against employees based on their illness. No employees shall have their job security threatened by their seeking of assistance for a substance abuse problem. The same consideration for referral and treatment that is afforded to other employees having nondrug/alcohol-related illnesses shall extend to them.

- Every effort shall be made to provide an early identification of a substance abuser, to work with and assist the employee in seeking and obtaining treatment without undue delay.
- Early identification of the substance abuser shall be based upon job performance and related criteria, as well as resulting impairment on the job from the job activities. The supervisor of the employee shall bring such information to the attention of the designated representative for further evaluation. An employee who voluntarily seeks treatment for a substance abuse problem, which requires a leave of absence consistent for treatment, shall be granted such leave of absence consistent with Company policies and procedures for other medical leaves of absence, and further shall be eligible for benefits under the specifications of the existing insurance plans.

Both applicants and employees have the right to refuse to undergo drug and alcohol testing. However, the consequences of refusal to promptly submit to such testing will be, in the case of an applicant, refusal to hire, and in the case of an employee, termination.

Nothing in this policy is construed to prohibit Western States Construction Company from fulfilling its goal of maintaining a safe and secure work environment for its employees or from invoking such disciplinary actions as may be deemed appropriate for actions of misconduct by virtue of their having arisen out of the use or abuse of alcohol or drugs or both.

Timekeeping

Federal and state laws require Western States Construction Company to keep an accurate record of time worked in order to calculate employee pay and benefits. Nonexempt employees should accurately record the time they begin and end their work, as well as the beginning and ending time of each meal period on their time card. They should also record the beginning and ending time of any split shift or departure from work for personal reasons. Overtime work must always be approved before it is performed.

Exempt employees are required to record any exceptions to their normal work schedule, i.e. time off, or other situations that need to be recorded for company or legal purposes. It is the employees' responsibility to sign or to certify the accuracy of all time recorded and request corrections or modifications. The supervisor should review and then initial the time record before submitting it for payroll processing. The employee should notify the Human Resources staff immediately if they feel a discrepancy has not been taken care of appropriately.

Work Schedule

Western States Construction Company's regular business office hours are from 8:00 a.m. to 5:00 p.m. Monday through Friday. Talk to your supervisor for your individual work schedule. Nonexempt employees working a full day will be scheduled for an unpaid meal break approximately half way through the day, operating requirements permitting. Paid rest breaks are normally taken midway through each half of the workday. At a minimum, a meal break is 30 minutes and a rest break is 10 minutes. The length and timing of each break should be scheduled to minimize disruption to work flow and service to the customer.

Part-time, nonexempt employees working five hours or less are not normally scheduled for a meal break and can waive the meal break if they are working no more than six hours. In all cases, the work schedule needs to be approved in advance by the supervisor or manager.

Workweek

For pay calculation purposes, the workweek begins at 12:01 a.m. Monday morning and ends the following Sunday at midnight.

Overtime

When operating requirements or other needs cannot be met during regular working hours, nonexempt employees may be scheduled to work overtime hours. When possible, advance notification of these mandatory assignments will be provided. All overtime work must receive the supervisor's prior authorization.

Overtime compensation is paid to all nonexempt employees in accordance with federal and state wage and hour regulations. Employees working under union agreement should refer to their labor agreement for any variation to these requirements.

Qualifying Hours: Only hours actually worked qualify for calculation of overtime premium pay. Holiday, vacation, sick, or any other leave of absence hours are not considered hours worked for the purpose of calculating overtime.

Exempt Employees: As stated previously, exempt employees do not qualify for overtime pay.

Personnel Files

Keeping your personnel file up-to-date can be important to you with regard to pay, deductions, benefits, and other matters. If you have a change in any personal information, please be sure to notify your supervisor or the Human Resources staff as soon as possible. Since the Company refers to your personnel file when we need to make decisions in connection with promotions or assignments, it is to your benefit to ensure that your personnel file is current.

You may review the information that is kept in your own personnel file, if you wish, and may, at your own expense, request and receive copies of all documents that you have signed. To view your personnel file, please submit a request to the Human Resources staff. As a reminder, removal of any personnel records is not allowed.

Garnishments

If the Company is instructed by the appropriate authority to garnish your pay, you will be informed of our obligation and asked to take whatever steps you can to resolve the garnishment without requiring our involvement. If the notice is not rescinded by the requesting authority, we will comply with the garnishment notice as directed.

Garnishments represent an additional burden on the Payroll Department. You are requested to deal with the issues that may lead to a garnishment to avoid involving the Payroll Department in your personal affairs.

Severance

Western States Construction Company at its sole discretion may provide severance pay to employees whose employment is terminated. Specifically excluded from benefits under this provision are employees who were hired as temporary employees for a specified period of time; were offered but refused to accept another suitable position with the organization; were terminated within six months of employment or introductory period.

Employee Compensation

It is the Company's desire to pay wages and salaries that are competitive with other employers in the marketplace in a way that will be motivational, fair and equitable, variable with individual and Company performance, and in compliance with all applicable statutory requirements. You are employed by Western States Construction Company and will be carried directly on the payroll. No person may be paid directly out of petty cash or any other such fund for work performed. The only exception to this policy is where a contract relationship exists with a bona fide contractor.

Compensation Reviews

Except as otherwise specified in the collective bargaining agreement, wage and salary increases are based on merit alone, not length of service or the cost of living. Any wage or salary increases will appear in the pay period ending after the dates they are granted.

Deductions

Western States Construction Company is required by law to make certain deductions from your paycheck each time one is prepared. Among these are your federal, state, and local income taxes and your contribution to Social Security as required by law. These deductions will be itemized on your check stub. The amount of the deductions may depend on your earnings and on the information you furnish on your W-4 form regarding the number of dependents or exemptions you claim.

It may be possible to make additional deductions from your paycheck, such as contributions to a 401(k), contributions to a medical or dependent care plan, or to deposit your paycheck directly into your savings or checking account at a participating bank. Contact the Human Resources staff for details and the necessary authorization forms.

Errors in Pay

Every effort is made to avoid errors in your paycheck. If you believe an error has been made, tell your manager immediately. He or she will take the necessary steps to research the problem and to assure that any necessary correction is made properly and promptly.

Lunch Periods and Rest Breaks

Your work schedule will be determined between you and your supervisor. Nonexempt employees must take one 10-minute break for each four (4)-hour period worked. Please arrange with your manager to take these breaks at a time of convenience for you, your coworkers, and your clients.

Overtime Pay

From time to time it may be necessary for you to perform overtime work in order to complete a job on time. We hope to give you advance notice of overtime. However, because of the nature of your work and our reputation for service, our employees stand ready to work overtime when requested or required. Overtime is *not* an option. When a manager requests you to work overtime, you are expected to cooperate as a condition of your employment.

If you are a nonexempt employee and you perform overtime work, you will be paid one and one-half ($1\frac{1}{2}$) times your regular hourly wage for any time over forty (40) hours per week, or eight (8) hours in one day. If you are scheduled for four 10-hour shifts, overtime will be paid for any time over forty (40) hours per week, or ten (10) hours in one day. If, during that week, you were away from the job because of a job-related injury, paid time off for bereavement, jury duty, vacation, or paid sick time, those hours will not be counted as hours worked for the purpose of computing eligibility for overtime pay.

Full-time "nonexempt" employees who work on a Company holiday will receive their normal wages for the paid holiday, plus they will be paid one and one-half ($1\frac{1}{2}$) times their regular hourly rate of pay for hours worked on the Company holiday regardless of the number of hours they work that workweek.

All overtime work by a nonexempt employee must be approved in advance by your manager. Because unauthorized overtime is against Company policy, employees who work unauthorized overtime are subject to disciplinary action, up to and including dismissal.

Exempt employees are paid a fixed salary that is intended to cover all of the compensation to which they are entitled. Because they are exempt, such employees are not entitled to additional compensation for extra hours of work or time off in lieu of additional compensation. The Company does not maintain any compensatory time-off plan or arrangement.

Pay Periods and Hours

Our payroll workweek begins on Monday at 12:01 a.m. and ends on Sunday at 12:00 midnight. Employees are paid on the 10th^h and 25th of each month. Pay periods are as follows:

Days worked	Pay Day
1st through 15th	25th of same month
16th through end of month	10th of following month

If payday falls on a Saturday, payday will be Friday. If payday falls on a Sunday, payday will be Monday. Paydays that fall on scheduled holidays will be paid at the discretion of management.

We will not release your paycheck to anyone other than you, except with your written authorization. Paychecks are mailed to your home address or directly deposited to the checking or savings account you specify. The distribution method is your choice and may be changed at any time. To make a change in the method that your paycheck is distributed to you, or to inquire about other options, see your manager or the Human Resources staff.

Performance Reviews

As a new employee, your progress will be continually reviewed during the Probationary Period. A formal evaluation of your progress will be made at the end of your Probationary Period. At that time, a decision will be made to offer you regular, full-time employment, extend the Probationary Period, or terminate your employment.

If you are offered full-time employment, your supervisor will formally review your job performance once annually thereafter. Your performance review will be based on the job requirements specified in your job description, or advice given to you by Management to increase your efficiency, or on reasonable requests of your manager. Each formal appraisal will be shown to you and discussed in detail. The formal appraisals are also reviewed by Management and, when final, become a part of your official personnel file.

The work of all employees is also appraised informally by your manager on a continuing basis. This continuous appraisal, as well as your more formal evaluations, is based on the job requirements specified in your position description, on special objectives added during the period since your last evaluation, and on other reasonable requests made by your manager. Your manager may suggest ways that you can improve the quality of your work or qualify for promotions. In fact, it is part of your manager's job responsibility to help you grow in your job.

Time Cards/Records

By law, we are obligated to keep accurate records of the time worked by nonexempt employees. This is done by time clock, time cards, or other written documentation. You are responsible for your time card. Remember to record your time and have your manager initial it. If you forget to punch in or if you make an error on your card, your manager must make the correction and you and your manager must initial the correction. You must also record your own time whenever you leave the work site for any reason other than for Company business.

Time cards are due on or before the designated time and date unless prior arrangements have been made with Payroll. Failure to submit time sheet(s) or failure to submit them in a timely manner may result in disciplinary action, up to and including dismissal. All hours worked by a nonexempt employee must be recorded on a daily basis and reported by the employee under this policy. By signing the time card, you agree to the accuracy and truthfulness of the hours worked and consent to the Company's payment of those hours.

No one may record hours worked on another's time card. Tampering with another's time card is cause for disciplinary action, including possible dismissal, of both employees. Do not alter another person's record, or influence anyone else to alter your record for you. In the event of an error in recording your time, please report the matter to your manager immediately.

Employee Benefits

Although it is our intention to provide a wide array of good benefits, Western States Construction Company reserves the right to modify, amend, or terminate its welfare and retirement benefits as they apply to all current, former, and retired employees. The administrator of each benefit plan has the discretionary authority to determine eligibility for benefits and to construe the plan's terms.

Eligibility for Benefits

If you are a full-time employee, you will enjoy all of the benefits described in this manual as soon as you meet the eligibility requirements for each particular benefit. If you are a part-time employee, you will enjoy only those benefits required by law to be afforded to you, provided that you meet the minimum requirements set forth by law and in the benefit plan(s). No benefits are available to you during your Introductory Period, except as otherwise provided by law or indicated in this manual. Temporary or casual employees are not eligible for benefits.

The specific plan documents of each welfare or retirement plan govern the rules of eligibility and your legal rights. The Human Resources staff is available to answer questions concerning benefits.

Paid Leaves of Absence

Holidays Only full-time employees are eligible for holiday pay. You are not eligible to receive holiday pay during your Introductory Period. The following holidays are recognized as paid holidays:

New Year's Day
Memorial Day
Independence Day
Labor Day
Thanksgiving Day
Day after Thanksgiving Day
Christmas Day

You may take time off to observe your religious holidays. If available, a full day of unused vacation may be used for this purpose, otherwise the time off is without pay. You must notify your manager at least five (5) working days in advance. We schedule all national holidays on the day designated by common business practice. If a holiday occurs during your scheduled vacation, you are permitted to take an extra day of vacation. In order to qualify for holiday pay, you must work the scheduled workday immediately before and after the holiday. Only excused absences will be considered exceptions to this policy. You are not eligible to receive holiday pay when you are on an unpaid leave of absence or disciplinary suspension.

Vacation Vacation is a time for you to rest, relax, and pursue special interests. Western States Construction Company provides paid vacation as one of the many ways in which we show our appreciation for your loyalty and continued service. Full-time employees are eligible to accrue vacation for each calendar month of service from their date of hire. The vacation accrual rate is based on your length of employment, as shown in the chart below.

Years of Employment	Total Accrual/Year (Days)
Less than one (1) year	None
Over 1 but less than two (2) years	5 Days
Two (2) or more years	10 Days

Every effort will be made to grant vacation leave at the time you desire. However, vacations cannot interfere with your team's operation and therefore your manager must approve all vacation requests. Vacation time that is not taken does not accumulate from year to year, except under unusual circumstances where an employee is unable to take vacation due to pressing work requirements. If you leave the Company, you will be paid for all vacation time you have earned but not taken. You will not be paid for vacation you are earning for the next year. Consult your supervisor if you have any questions about the vacation policy.

Sick Leave All full-time employees who have completed their Probationary Period are eligible to take up to one half day of sick leave with pay each month. Sick leave is intended to be used only when actually required to recover from illness or injury; sick leave is not for "personal" absences. Time off for medical and dental appointments will be treated as sick leave.

If you take time off because of illness or personal injury, "proof of illness' may be required before sick time will be paid. If you are absent longer than three (3) business days due to illness, medical evidence of your illness and/or medical certification of your fitness to return to work by a physician satisfactory to the Company will be required before sick pay will be given. As sick leave benefit is to provide for your well-being during employment, it is not payable at termination of employment, retirement, or death.

Funeral (Bereavement) Leave You are entitled to take up to three (3) workdays with pay to attend the funeral and take care of personal matters related to the death of a member of your immediate family. For the purposes of this policy, your immediate family includes your grandparent, parent, spouse, child, grandchild, and siblings, or your spouse's grandparent, parent, child, grandchild, and siblings. Employees may also request up to one (1) day of unpaid leave to attend the funeral of a non-immediate family member. If you prefer, a day of earned vacation may be used for this purpose.

Unpaid Leaves of Absence

Occasionally, for medical, personal, or other reasons, you may need to be temporarily released from the duties of your job without pay. Under certain circumstances, you may be eligible for an unpaid leave of absence. There are several types of unpaid leaves available based on eligibility.

Educational Leave of Absence An educational leave of absence may be approved, at the sole discretion of the Company, if the desired curriculum is of mutual benefit to you and to Western States Construction Company. Apply in the same manner as you would for a personal leave of absence.

Election Day The Company considers voting an employee's privilege and duty. Our scheduled work hours should permit adequate time to vote before or after normal working hours. If you are unable to vote during nonworking hours because of your schedule, please contact your manager to arrange for time off to vote.

Federal Family Medical Leave Act (FMLA) In compliance with Federal law, the Company provides leaves of absence to eligible employees who want to take time off to fulfill family obligations relating directly to childbirth, adoption, or placement of a foster child; to care for a child, spouse, or parent with a serious health condition; or for a serious health condition that makes you unable to perform your job.

Employees are eligible upon completion of 12 months of service and 1,250 hours worked in the preceding 12-month period. Eligible employees should make requests for leave at least 30 days in advance of foreseeable events and as soon as possible for unforeseeable events. Reasonable advance notice is required, even in emergencies.

Employees requesting leave related to their own serious health condition or to the serious health condition of a child, spouse, or parent will be required to have the health care provider complete a certification which verifies the need for leave, its beginning and expected end date, and the estimated time required. All employees on leave need to keep in contact with the Human Resources staff regarding their status, and recertifications of the need for leave may be requested at reasonable intervals. Under certain circumstances, the Company may request a second or third medical opinion, at its expense, relating to an employee's medical certification of the need for a leave.

Employees are eligible for up to a maximum of 12 weeks of leave within the specified 12-month period. Employees using leave under this policy must first use any accrued paid vacation time, which time will count toward the 12-week maximum. Married employee couples are restricted to a combined total of 12 weeks of leave within the 12-month period for childbirth, adoption, or placement of a foster child; or to care for a parent with a serious health condition.

Western States Construction Company will maintain group health insurance coverage for employees on family and medical leave for up to a maximum of twelve (12) workweeks if such insurance was provided before the leave was taken, and on the same terms as if the employee had continued to work. In some instances, the Company may recover premiums it paid to maintain health coverage for employees who fail to return to work following family and medical leave. Benefit accruals, such as vacation or holiday benefits, will be suspended during the leave and will resume upon return to active employment.

When leave ends, employees will be required to submit to a fitness for duty certification prior to returning to work. If the employee returns to work after using no more than 12 weeks of leave allowed by this policy, the employee will be reinstated to the same position, if it is available, or to an equivalent position for which the employee is qualified. If an employee fails to report to work promptly at the end of the approved leave period, the Company will assume that the employee has resigned.

Pregnancy Leave If you work in Washington and take leave for pregnancy, you are entitled under Washington state law to unpaid leave for the full period of time you are sick or temporarily disabled because of pregnancy or childbirth, even if you do not qualify for leave under the Federal law described above. Also, if you qualify, Washington law may allow you to take up to an additional 12 weeks of unpaid leave for child care after you are physically able to return to work.

Jury Duty It is your civic duty as a citizen to report for jury duty whenever called. If you are called for jury duty, we will permit you to take the necessary time off to serve. However, Western States Construction Company does not compensate employees for time spent serving on jury duty. Exempt employees who work any part of the workweek while also serving on jury duty will be paid for the entire workweek. You must notify your manager within forty-eight (48) hours of receipt of the jury summons.

Military Leave of Absence Employees who serve in the United States military organizations or state National Guard may take the necessary time off without pay to fulfill this obligation, and will retain all their legal rights for continued employment under existing laws. These employees may apply accrued vacation time to the leave if they wish; however, they are not obligated to do so. You are expected to notify your supervisor as soon as you are aware of the dates you will be on duty so that arrangements can be made for replacement during this absence.

Insurance Coverage

General Information Western States Construction Company has established a variety of benefit plans designed to assist you and your eligible dependents in meeting the financial burdens that can result from illness and disability, and to help you plan for retirement. This portion of the handbook contains a general description of the benefits to which you may be eligible. Please understand that this general explanation is not intended to, and does not, provide you with all the details of these benefits. Therefore, this manual does not change or otherwise interpret the terms of the official plan documents. Your rights can be determined only by referring to the full text of the official plan documents, which are available for your examination from the Human Resources staff. To the extent that any of the information contained in this manual is inconsistent with the official plan documents, the provisions of the official documents will govern in all cases. For more complete information regarding any of our benefit programs, please refer to the policy certificates or the Summary Plan Description(s).

Group Insurance Western States Construction Company employees deserve a complete benefits package—a package that demonstrates concern for the welfare of our workers, your dependents and your eventual survivors. We want you to be relieved of the worry and fear associated with possible financial disaster from unexpected or unplanned events.

Eligibility The Company benefits package is available to all full-time employees who work at least 30 hours per week. Full-time, salaried employees are eligible to participate in the program on their date of hire. Full-time, hourly employees are eligible to participate in the program on the first day of the month following 90 days of continuous service. Employees may choose to cover eligible dependents. For more information about eligibility, please refer to the benefits enrollment materials.

The core benefits package includes the following group insurance programs:

- Basic Life and Accidental Death and Disability Insurance
- Short-Term Disability Insurance
- Long-Term Disability Insurance
- A choice of three medical options

Basic Life and Accidental Death and Disability Insurance The basic life portion of the plan pays benefits if you die. The accidental death and disability portion of the plan pays benefits if you die or are seriously injured as the result of an accident. If you die as a result of an accident, your beneficiary will receive benefits from both the basic life and the accidental death and disability portions of the plan. This plan provides $75,000 of life insurance and $75,000 of accidental death and disability coverage.

Short-Term Disability Insurance Short-term disability is a core benefit that provides income replacement benefits in the event you become disabled or are unable to work. The plan provides a benefit equal to 60 percent of your base weekly salary, up to $600 per week to partially offset your loss of income. Short-term disability benefits begin on the first day of an accident or the eighth day of an illness. Eligible employees continue to receive weekly benefits for up to 26 weeks while disabled.

Long-Term Disability Insurance If you become totally disabled and cannot perform the main duties of your regular job, you can receive long-term disability benefits for up to two years. If the disability continues for more than two years and you are deemed to be unable to perform any occupation that you are reasonably suited to do by training, education, and experience, your monthly benefit can continue up to age 65. The long-term disability benefits pay you 60 percent of your base monthly salary, up to $10,000 per month while disabled.

Medical Coverage Options For most people, medical benefits are the central focus of a benefits package. So that you can select the coverage that is right for you and your family, Western States Construction Company offers three medical options. The choices for medical coverage are based on your home zip code. Depending on where you live, you can enroll in either of the following:

- Cascade Health Exclusive Provider Program (HMO)
- Cascade Health Preferred Provider Program (PPO)

Please refer to enrollment materials for a more detailed description of the medical plan options.

Termination of Insurance Your insurance will terminate when the insurance policy terminates, when you fail to make an agreed contribution to the premium when due, when you cease to be eligible for coverage under the terms of our group insurance program, or when you cease to be employed as a regular full-time employee eligible for the insurance.

Western States Construction Company may, by continuing to pay the premium, keep your insurance in effect for a brief period if you cease to be an eligible employee for any reason other than resignation, dismissal, or failure to meet the terms of eligibility of our group insurance program.

You and your eligible dependents may have the right to continued coverage under COBRA (Consolidated Omnibus Budget Reconciliation Act) for a limited period of time at your or their own expense. (This does not affect the conversion privilege as stated in the insurance policy.)

Consolidated Omnibus Budget Reconciliation Act In accordance with Federal law, most employers sponsoring group medical/dental plans are required to offer employees and their families the opportunity for a temporary extension of medical/dental coverage, called continuation coverage, at group rates in certain instances where coverage under the plan would otherwise end. This notice is intended to inform you, in a summary fashion, of your rights and obligations under the continuation coverage provisions of the law.

As an employee of Western States Construction Company you may have the right to choose this continuation coverage if you lose your group medical/dental coverage because of a reduction in your hours of employment or the termination of your employment for any reason other than gross misconduct on your part. The spouse or dependent child of an employee has the right to choose the continuation coverage under the Company's group medical/dental insurance plan under certain qualifying events.

Under the law, the employee or family member (including divorced spouses or dependent children) has the responsibility to inform the Company's Plan Administrator of a divorce, legal separation, or a child losing dependent status under the Company's group medical/dental plan.

The Company has the responsibility to notify the Plan Administrator of the employee's death, termination of employment, reduction in hours, or Medicare entitlement. When the Plan Administrator is notified that one of these events has happened, it will in turn notify you that you have the right to choose continuation coverage. Under the law, you have at least 60 days from the date you lose coverage because of the events described above to inform the Plan Administrator that you want continuation coverage. *If you do not choose continuation coverage, your group medical/dental insurance coverage will end.*

If you choose continuation coverage, Western States Construction Company is required to give you coverage, which, as of the time coverage is being provided, is identical to the coverage provided under the plan to similarly situated employees or family members. The law requires that you be afforded the opportunity to maintain continuation coverage for 36 months unless you lost group medical/dental coverage because of termination of employment or reduction in hours. In that case, the required continuation coverage period is 18 months. However, the law also provides that your continuation coverage may be cut short for other reasons. See the Plan Administrator for details.

Under the law, you are responsible for the entire premium for your continuation coverage. The law also says that, at the end of the 18-month or 36-month continuation period, you must be allowed to enroll in an individual conversion medical plan provided under the Company's medical insurance plan at your own expense. This does not apply to dental insurance plans.

Additional information regarding coverage and cost as well as a complete copy of the COBRA law may be obtained from the Human Resources staff. The Company will charge an administration fee of up to 2 percent of your monthly premium for processing paperwork and administering payments.

Government Required Coverage

Social Security The United States Government operates a system of contributory insurance known as Social Security. As a wage earner, you are required by law to contribute a set amount of your weekly wages to the trust fund from which benefits are paid. As your employer, Western States Construction Company is required to deduct this amount from each paycheck you receive. In addition, the Company matches your contribution dollar for dollar, thereby paying one-half of the cost of your Social Security benefits.

Unemployment Compensation Western States Construction Company pays a percentage of its payroll to the Unemployment Compensation Fund according to the Company's employment history. If you become unemployed, you may be eligible for unemployment compensation, under certain conditions, for a limited period of time. Unemployment compensation provides temporary income for workers who have lost their jobs. To be eligible you must have earned a certain amount and you must be willing and able to work. You should apply for benefits through your local State Unemployment Office as soon as possible.

Worker's Compensation Western States Construction Company provides a comprehensive worker's compensation program at no cost to employees. This program covers an injury or illness sustained in the course of employment that requires medical, surgical, or hospital treatment. Subject to applicable legal requirements, worker's compensation insurance provides benefits after a short waiting period or, if the employee is hospitalized, immediately.

Employees who sustain work-related injuries or illness must inform their manager immediately. No matter how minor an on-the-job injury may appear, it is important that it be reported immediately. This will enable an eligible employee to qualify for coverage as soon as possible.

Retirement

401(k) Plan Saving for your retirement is one of the most important financial decisions you can make. Western States Construction Company is pleased to offer a solid salary savings benefit that will help you prepare for your financial future. We offer a salary savings plan with provisions for employee and employer contributions. It is the type of plan commonly referred to as a 401(k) plan. You become eligible to participate after you have completed ninety days and have attained 18 years of age .

As an eligible employee, you are able to make contributions to the plan through payroll deduction in amounts from 1 percent to 15 percent of your before-tax salary, up to the IRS maximum. In addition, rollovers from another employer's qualified retirement plan are accepted. Western States Construction Company will help you by making periodic, discretionary contributions to the plan. There are several great reasons to take advantage of the plan.

- Your contributions are deducted automatically from your paycheck, making it easier to stick to a regular savings schedule.
- Your contributions are deducted from your pay on a *pretax* basis. This means that all Federal and state taxes are calculated on what you make after your 401(k) contribution is deducted.
- Earnings on your contributions grow on a tax-deferred basis. That is, you don't have to pay taxes on the growth in your account until you start taking the money out.
- Western States Construction Company wants to help you build your retirement savings foundation. That's why we will make a 50 percent matching contribution on the first 6 percent of deferrals, up to a maximum of $2,000.

We hope you will take advantage of this part of your benefits package. More information including investment options and restrictions will be made available when you are eligible to enroll. For more information on retirement and/or savings contact the Human Resources staff.

Company Standards and Rules

Personal Information and Personnel Records

It is Western States Construction Company's policy to verify current and past employment of its employees. We normally limit the information to your dates of employment and your job title. However, in cases where rate of pay or earnings history (as in the case of satisfying a loan application) is requested, we will provide financial information with your authorizing signature on the request. In certain instances, the Company is obligated by law to provide additional employment information when requested.

You have a responsibility to keep the Company advised of any changes to key information contained in your personnel records. Please communicate to Human Resources any changes in any of the following:

- Contact information (such as telephone numbers)
- Address
- Emergency contacts
- Insurance information (such as beneficiaries or dependents)

Employment Applications

The Company relies upon the accuracy of information contained in your employment application or any other information submitted for employment consideration, as well as the accuracy of other data presented throughout the hiring process and employment. Any misrepresentations, falsifications, or material omissions in any of this information or data may result in the Company's exclusion of the individual from further consideration for employment or, if the person has been hired, termination of employment.

Your Conduct

Western States Construction Company is committed to establishing a professional and productive work environment. You are expected to support that goal as a way of contributing to your own success, the success of your coworkers and, ultimately, the Company's success. The following rules of conduct are intended to ensure that employees consistently demonstrate to fellow team members, customers, and visitors, professional and courteous personal conduct. In addition, the rules also promote a safe work environment, which is always a primary concern at the Company.

Failure to comply with these standards will result in disciplinary action, up to and including termination. The Company reserves the right to proceed directly to a written warning or to termination, without resort to prior disciplinary steps, when the company deems such action appropriate. It is not possible to list all the forms of behavior that are considered unacceptable in the workplace. The following are examples of infractions of rules of conduct that may result in disciplinary action, up to and including termination of employment:

- Failure or refusal to carry out job assignments
- Unauthorized release of Company confidential information
- Falsification of work, personnel, or other Company records
- Dishonesty or theft
- Insubordination
- Discrimination or unlawful harassment
- Deliberate damage or conversion of Company property
- Fighting in the workplace
- Unauthorized possession of firearms or other dangerous weapons at the workplace
- Violation of safety rules
- Negligence or improper conduct leading to damage of employer-owned or customer-owned property
- Sexual or other unlawful or unwelcome harassment
- Inappropriate use of computers, computer files, the e-mail system, software, telephone, cell phones, voice mail, and any other voice or electronic communication systems furnished to employees
- Excessive absenteeism or any absence without notice
- Violation of personnel policies
- Unsatisfactory performance or conduct
- Sleeping on the job

As stated above, an all-inclusive list of infractions is not possible and the ones given above are provided as examples. You are encouraged to discuss this policy with your supervisor. Employment with the Western States Construction Company is at the mutual consent of the Company and the employee, and either party may terminate that relationship at any time, with or without cause, and with or without advance notice.

Media Inquiries

You are expected to refrain from talking about the Company business to members of the press or broadcast organizations. To safeguard the accuracy of any publicity regarding the Company, only the President is authorized to release information to the press. If a reporter from any publication, radio, or television station requests information from you, politely take down the information requested. In addition, ask the reporter for his or her name and organization. Tell the caller you will refer the request to the President.

Solicitations

Persons not employed by Western States Construction Company may not solicit or distribute literature on Company property for any purpose at any time. This prohibition includes, but is not limited to, charitable solicitors, sales people, surveyors or questionnaires, political messages, and any other form of solicitation, distribution, or leaflets. Employees may not solicit verbally for any purpose during actual working time. "Actual working time" means no time during which you are required to perform work duties and does not include break times, meal times, and before and after work. Employees may not distribute literature for any purpose during working time or in working areas.

Attendance and Punctuality

Each person is expected to be punctual and maintain good attendance. While none of us can completely avoid absences due to personal illness or other causes, time away from work creates problems for our customers and fellow employees. For this reason employees are expected to commit to working the scheduled hours established with their supervisor. This includes arriving at work on time and being available during the day as specified by the job and the supervisor.

Unscheduled absences from the job require adequate and advance notification by the employee. If you are unable to report for work, you are expected to contact your supervisor personally as soon as reasonably possible but always before your scheduled starting time. You will be expected to communicate the expected duration of your absence and an explanation for the absence. An employee who fails to report to work and fails to call in for a period of more than three days will be assumed to have resigned voluntarily from the job.

Personal Appearance

Dress, grooming, and personal cleanliness standards contribute to the morale of all employees and affect the business image the Company presents to the community. During business hours, employees are expected to present a clean and neat appearance and dress according to the requirements of their positions. Consult your supervisor if you have questions as to what constitutes appropriate attire.

Safety

Western States Construction Company believes that all incidents and injuries are preventable, and demonstrates this belief through the daily business activities of the corporation. Our employees are the Company's greatest resource. In order to protect this resource, we need to continue to instill a safety culture that permeates every level of the Company and every Company work site.

To assist in providing a workplace free from recognized hazards, each supervisor will receive a copy of our corporate safety policies and procedures manual as well as a jobsite injury and illness prevention program. These manuals are available through the Safety Department. These programs are based on a sincere desire to eliminate injuries and illnesses, damage to equipment and property, as well as to protect the public, customers, and vendors whenever and wherever they may be affected by Company work.

Management Commitment: All supervisors are responsible for ensuring that employee involvement and participation in the safety process is maximized as appropriate for individual work sites. Supervisors are responsible for supporting and encouraging their employees to participate in activities that contribute to the continuous improvement of safety.

Employee Involvement and Participation: There is an important role in this program for each employee, and everyone is expected to join together to make the Company a successful, accident-free, and healthy place to work. Employees are responsible for participating in safety activities such as safety meetings, safety committees, correcting a hazardous situation, or completing the job hazard analysis. Remember, every Company employee deserves to go home in the same condition in which they arrived at the workplace. Let's make this a reality at the Company.

In case of accidents that result in injury, regardless of how insignificant the injury may appear, employees should immediately notify the Safety Director or the appropriate supervisor. Such reports are necessary to comply with laws and initiate insurance and worker's compensation benefits procedures.

Security Inspections

Western States Construction Company wishes to maintain a work environment that is free from illegal drugs, alcohol, firearms, explosives, or other improper materials. To this end, the Company prohibits the possession, transfer, sale, or use of such materials on its premises. The Company requires the cooperation of all employees in administering this policy.

Desks, lockers, and other storage devices may be provided for the convenience of employees but remain the sole property of the Company. Accordingly, any appointed agent or representative of the Company can inspect them, as well as any articles found within them, at any time, either with or without prior notice.

The Company likewise wishes to discourage theft or unauthorized possession of the property of the employees, the Company, the visitors, and the customers. To facilitate enforcement of this policy, the Company or its representative may inspect not only desks and lockers but also persons entering and/or leaving the premises and any packages or other belongings. Any employee who wishes to avoid inspection of any articles or materials should not bring such items onto the Company's premises.

Workplace Security

Western States Construction Company is committed to preventing workplace violence and to maintaining a safe work environment. Given the increasing violence in society in general, we have adopted the following guidelines to deal with intimidation, harassment, or other threats of (or actual) violence that may occur during business hours or on its premises.

All employees, including supervisors, managers, and temporary employees, should be treated with courtesy and respect at all times. Employees are expected to refrain from fighting, "horseplay," or other conduct that may be dangerous to others. Firearms, weapons, and other dangerous or hazardous devices or substances are prohibited on the premises of the Company or its jobsites without proper authorization.

Conduct that threatens, intimidates, or coerces another employee, a customer, or a member of the public at any time, including off-duty periods, will not be tolerated. All verbal and/or physical threats of (or actual) violence, both direct and indirect, should be reported as soon as possible to your immediate supervisor or any other member of management. This includes threats by employees, as well as threats by customers, vendors, solicitors, or other members of the public. When reporting a threat of violence, you should be as specific and detailed as possible.

Visitors in the Workplace

To provide for the safety and security of employees and the facilities at the Company, only authorized visitors are allowed in the workplace. Restricting unauthorized visitors helps maintain safety standards, protects against theft, ensures security of equipment, protects confidential information, safeguards employee welfare, and avoids potential distractions and disturbances.

All visitors should enter the Company at the reception area. Authorized visitors will receive directions or be escorted to their destination. Employees and/or their visitors are responsible for the conduct and safety of their visitors. If an unauthorized individual is observed on the Company's premises, employees should immediately notify their supervisor or direct the individual to the reception area.

Smoking

In keeping with the Company's intent to provide a safe and healthful work environment, smoking is prohibited throughout the workplace. The company provides designated smoking areas. Employees should ask their supervisors where the designated smoking area(s) are located. This policy applies equally to all employees, customers, and visitors.

Glossary

Accounting the process of collecting, analyzing, classifying, and accumulating historical financial data in categories that will reflect the financial condition of a company's operation.

Accounts payable debts owed to suppliers and subcontractors or for the purchase of any good or service.

Accounts receivable claims against others for work performed or assets sold for which payment has not been received.

Accrual method of accounting an accounting system that recognizes revenues when they are earned and expenses when they are incurred, regardless of when the cash transaction takes place.

Action plan an assignment of specific actions to an individual, an allocation of resources, and an establishment of a specific date by which the actions are to be completed.

Additional named insured additional parties provided coverage by an insurance policy.

Assets things of value that are owned by a company.

Automobile insurance insurance that provides coverage for financial loss due to the loss of or physical damage to vehicles and for liability for bodily damage to others or damage to the property owned by others caused by the operation of a vehicle.

Balance sheet a financial statement that represents the financial condition of a company as of the date of the balance sheet in terms of assets, liabilities, and owners' equity.

Bid bond a guarantee that a successful bidder will furnish a performance bond and other contractual requirements and enter into a construction contract for the price contained in its bid.

Boiler and machinery insurance insurance that covers financial loss due to mechanical breakdown of boilers, electrical equipment, pumps, motors, compressors, or air-conditioning equipment as well mechanical breakdown of the covered equipment.

Bonding agent (also known as surety bond producer) a representative of a construction company who identifies the most advantageous bonding arrangement with a surety.

Book value the current value of a depreciable asset, which is the acquisition price less accumulated depreciation.

Bridge insurance insurance that provides builder's risk type of coverage to bridge projects that may be excluded from coverage by conventional builder's risk policies.

Builder's risk insurance insurance that provides coverage for the cost of damages to the project under construction, to include temporary structures and any materials stored on the project site, but not the contractor's equipment.

Buy/sell agreement an agreement that specifies how the value of each partner's share of a partnership is to be determined and the right of the remaining partners to purchase the departing member's share.

Cash flow analysis an analysis of the income received and expenses incurred for a single project or for all the projects being undertaken by a company.

Cash method of accounting an accounting system that reports revenues and expenditures in the accounting period in which cash is received or disbursed, regardless of when the revenues were earned or the expenses incurred.

Chart of accounts a listing of numerical codes used to classify data captured in a company's accounting system.

Claims-based liability insurance insurance that provides coverage only for claims made during the policy period, irrespective of when the insured occurrence took place.

Completed-contract accounting method an accounting system that does not take credit for any profit until a construction project has been completed and the contract closed out.

Controlling monitoring activities to ensure that they are being accomplished as planned and correcting any significant deviations.

Core competencies the collective skills, knowledge, and experience possessed by the employees of the company.

Core values a statement of company values that provides moral guidance for company employees.

Corporation a legal entity formed under state law by a certain number of stockholders filing a certificate of articles of incorporation with an appropriate official of the state government and which operates under a corporate name.

Current assets non-depreciable assets such as cash, accounts receivable, and inventory.

Current liabilities debts that are expected to be paid within a year.

Current ratio a measure of the liquidity of a company that is determined by dividing the current assets by the current liabilities.

Debt analysis analysis of how much debt a company is using to finance its operations.

Debt-to-equity ratio a measure of the leverage of a company that is determined by dividing the total liabilities by the owners' equity.

Demographic environment the age distribution of the population, its ethnic composition, the regional changes in population growth, the employment rates, education levels, and the availability of prospective new employees.

Directors and officers liability insurance insurance that provides company directors and officers protection from personal liability and financial loss arising out of acts committed or allegedly committed in their capacity as company directors or officers.

Divisional organization an organizational structure that contains multiple functional elements or divisions that focus on different markets.

Economic environment the economic factors that affect the customers' ability to purchase construction services and the construction company's cost of doing business.

Employment practices liability insurance insurance that provides protection for a company against claims related to employment issues made by employees, former employees, or potential employees.

Equal employment opportunity equal access to jobs and benefits and services for all employees and prospective employees.

Equipment floater insurance insurance the provides coverage for financial loss due to loss of or damage to construction equipment, temporary structures used to support construction operations, and materials and supplies that will not be incorporated into a construction project.

Errors and omissions insurance professional liability insurance that covers professional negligence of design professionals.

Ethics moral standards used by people in making personal and business decisions.

Experience modification ratio (EMR) a multiplier that is applied to the base premium rate to determine the rate that a company must pay for workers' compensation insurance. The EMR is based on the company's claim history in the oldest three of the past four years.

External assessment an analysis of forecast opportunities for future successes and threats that the company may face.

Extranets Web-based networks accessible only to company employees and business partners.

Financial statements statements produced by accountants that provide a picture of the financial condition of a company.

Firewalls hardware security devices with special software that are placed between the Internet and company networks.

Fixed assets depreciable assets that are retained for longer than one year.

Floating marine equipment insurance insurance that provides coverage for financial loss due to the loss of or physical damage to floating equipment and for liability for bodily damage to others or damage to the property owned by others caused by the operation of floating equipment.

Functional organization an organizational structure that contains separate elements that perform specialized functional activities.

General ledger system the primary accounting system of a company that is used by an accountant to generate financial statements.

Gross profit total sales less the cost of the sales.

Hardware   input devices, output devices, storage devices, central processing units, random-access memory, and connecting devices used to create information management systems.

Income statement   a financial statement that summarizes the profitability of a company over a period of time.

Indemnity agreement   an agreement that provides that a construction company will pay back a surety for any losses the surety incurs in making good on the guarantee to the obligee.

Insurance agent   an individual who advises clients regarding the purchase of insurance policies from insurance companies.

Insurance policy   a two-party contract under which the insurance company promises, for a fee, to assume responsibility for specific losses or liabilities of the insured for a specified period of time.

Internal assessment   an analysis of the strengths and weaknesses of a company.

Internal audit   assessing the company's major functional areas to determine if they have adequate resources to perform their work activities and how well they perform their assigned tasks.

Internal customers   employees who receive support from various departments within a company.

Internal functional analysis   identifying and evaluating the company's resources, capabilities, and core competencies.

Internal marketing   informing company employees of company activities to ensure they understand their roles in maintaining the company's reputation and in marketing its services.

Intranets   Web-based networks accessible only to company employees.

Job description   a description of the major tasks and duties to be performed by someone who occupies the position.

Job specifications   the knowledge, the abilities, and the skills needed by a person in order to be able to perform the tasks and duties contained in a job description.

Joint venture   a business alliance of two or more business enterprises.

Key-person insurance   life insurance obtained by a company on company principals.

Labor and material bond (also known as payment bond)   a guarantee that the construction contractor will compensate any entity that provided labor or materials for the construction project.

Leading   creating a vision for the company, motivating subordinates, establishing company standards of behavior and ethics, directing others, selecting communication channels, and resolving conflicts.

Leverage   the use of borrowed funds to finance a company's operations.

Liabilities   the value of claims owed to creditors of a company.

Liability insurance   insurance that provides coverage for financial loss that results from injuries or from property losses sustained by third parties as a consequence of a construction firm's activities.

Limited liability company   a company owned by two or more people that is taxed like a partnership while limiting personal liability.

Limited partnership   a partnership formed with two categories of partners, general partners and limited partners. A general partner contributes resources to the partnership and performs a management function, while a limited partner contributes resources, but is not involved in the management of the partnership.

Liquidity analysis   analysis of the ability of a company to meet its current obligations when they become due.

Maintenance bond (also known as a warranty bond)   a guarantee that the construction contractor will return to a completed project and repair or replace any defective or inferior materials or workmanship during the specified warranty period.

Marketing   the process of retaining existing customers, identifying prospective new ones, and attracting new customers to consider the construction firm as its service provider.

Matrix organization   an organizational structure that contains both multifunctional teams and functional chiefs.

Mechanic's lien   a legal claim against a property imposed by subcontractors or suppliers who have not been compensated for their work or materials.

Mission statement   a description of why the company exists and what makes it different from other companies in the industry.

Net profit after taxes   net profit before taxes less taxes.

Net profit before taxes gross profit less operating expenses.

Obligee the party receiving the guarantee of performance by a surety bond.

Occurrence-based policy insurance that provides coverage regardless of when a claim is made provided the insured occurrence took place during the policy period.

Organization chart pictorial representation of a company's organizational structure.

Organizational design the selection of the organizational structure for the company and the determination of the number of people needed for each organizational element.

Organizing determining what tasks are to be done, who is to do them, how tasks are to be grouped, who reports to whom, and where decisions are made.

Owners' equity (also known as company net worth or capitalization) the net value of the company that is owned by the owners of the company. It is determined by subtracting the total liabilities from the total assets of the company.

Partnership an association of two or more persons to conduct business as co-owners.

Penal sum the maximum liability of a surety on a surety bond.

Percentage-of-completion accounting method an accounting system that estimates the percentage completion of each construction project at the end of an accounting period to determine the profit realized during the accounting period.

Performance bond a guarantee that the construction contractor will perform according to the terms of the construction contract.

Performance management the establishment of performance standards for each position and the creation of a performance appraisal system.

Planning defining company goals, establishing strategies to accomplish goals, and developing plans to coordinate the activities of organizational elements.

Political/legal environment government expenditures, laws, regulations, judicial decisions, and political forces.

Pollution liability insurance insurance that provides coverage for legal liability relating to pollution conditions.

Principal the purchaser of a surety bond, such as a construction company.

Process action teams teams of company employees knowledgeable of a specific company business practice that are charged with evaluating the process and recommending improvements.

Profitability analysis analysis of the ability of a company to generate profits from its operations.

Project indirect costs project costs that are not allocated to a specific element of work, such as project overhead (job supervision, project office, bonding, etc.).

Property insurance insurance that provides coverage for financial loss due to physical damage to a construction company's property.

Proprietorship company owned by one individual.

Quality council a steering committee composed of top managers within the company that is responsible for establishing total quality policy within the company.

Quality management board a committee of senior managers that is responsible for managing quality improvement activities within the company.

Quick ratio a measure of the liquidity of a company that is determined by subtracting the inventory from the current assets and dividing the result by the current liabilities.

Railroad liability insurance insurance that provides liability insurance to cover bodily injury and property damage that may result from work executed near railroads.

Ratio analysis analysis of the financial characteristics of a company at a single point in time.

Relevant market the specific sector or sectors of the overall construction market that are of interest to the construction company.

Retained earnings company earnings that have not been paid to company owners.

Return on assets ratio a measure of the profitability of a company that is determined by dividing the net profit after taxes by the total assets.

Return on equity a measure of the profitability of a company that is determined by dividing the net profit after taxes by the owners' equity.

Return on sales a measure of the profitability of a company that is determined by dividing the net profit after taxes by the total sales.

Sales contacting a specific potential customer and winning the construction contract.

Sociocultural environment cultural values, attitudes, behaviors, and opinions.

Software computer programs that allow the information management system to perform the desired functions.

Span of control number of people supervised by each manager.

Staffing includes the recruitment, selection, and retention of company employees.

Strategic marketing enhancing the company name recognition and its reputation.

Strategic planning a process that involves assessing company strengths and weaknesses; analyzing the external environment; forecasting the future; and selecting objectives, strategies, and action plans that provide focus and guide company leaders in their decision-making processes.

Subrogation an insurance policy clause that grants the insurance company the right of the insured to recover from the party that caused the loss.

Supply bond a guarantee to a construction company that a material supplier will comply with the terms of a purchase order or supply contract.

Surety bond a guarantee on the performance of the principal.

Tactical marketing targeting specific current or prospective customers.

Technological environment technological changes and trends that may impact construction operations or construction management.

Total quality management (TQM) a management philosophy that focuses on continuous process improvement and customer satisfaction.

Transportation floater insurance insurance that provides coverage for direct loss of or physical damage to property while in transit to or from a construction project.

Trend analysis analysis to determine how the financial characteristics of a company have changed over time.

Type S corporation a closely held corporation that chooses to be taxed as a partnership.

Umbrella excess liability insurance insurance that provides high limits of liability coverage when the coverage provided by other liability insurance policies has been exhausted.

Value chain analysis a systematic approach to examining all of the company's functional activities and assessing how well they create customer value.

Vision a statement of what the company aspires to become in the future.

Worker's compensation insurance insurance that provides coverage for financial loss incurred by an employee as a result of work-related injury or disease.

Working capital the liquid assets that a company has to finance its future operations and it is determined by subtracting the current liabilities from the current assets.

Index

Accounting, 67
Accounting methods, 72
Accounting records, 71
Accounts payable, 75
Accounts receivable, 74
Accrual method of accounting, 72
Action plan, 98, 113, 114
Additional named insured, 36, 39, 40
Affirmative action, 160
Age discrimination, 161
All-risk insurance, 39, 41
Assets, 68, 74
Automobile insurance, 41

Balance sheet, 73–75
Bid bond, 54
Boiler and machinery insurance, 40
Bonding agent, 52, 59
Bonding capacity, 63
Bonding company. *See* Surety bond
Bond underwriting considerations, 58–64
Book value, 75
Bridge insurance, 40, 41
Builder's risk insurance, 39, 40
Business development, 117–133
Business insurance, 48
Business organization, 16–23
Business risk, 2–4, 34–36
Business strategies, 111, 112
Business volume, 2, 5
Buying cycle, 127, 128
Buy/sell agreement, 19, 48

Capital assets, 86
Capitalization. *See* Owners' equity
Cash flow analysis, 6, 81–84
Cash method of accounting, 72
Causes of business failure, 2–4
Chart of accounts, 68–70
Claims-based liability insurance, 44
Commercial general liability insurance, 42, 43
Communicative skills, 8
Company culture, 7, 137–139

Company Web sites, 142, 174, 175
Compensation, 156
Competitive strategies, 112, 113
Competitor analysis, 103, 125, 126
Completed-contract accounting method, 72, 73
Conceptual skills, 8
Contractor questionnaire, 60–63
Controlling, 7
Core competencies, 93, 105
Core values, 7, 12
Corporation, 17, 20–22
Current assets, 74
Current liabilities, 75
Current ratio, 76, 77
Customer analysis, 102
Customer satisfaction assessment, 124, 125

Debt analysis, 80
Debt-to-equity ratio, 80
Debt management, 3
Decision-making processes, 138
Demand assessment, 123, 124
Demographic environment, 101
Directors and officers liability insurance, 46
Disabled workers, 161
Discrimination, 161–165
  age discrimination, 161
  disabled workers, 161
  equal employment opportunity, 162, 163
  racial and ethnic discrimination, 163, 164
  religious protection, 164, 165
Divisional organization, 25, 26
Division of labor, 28
Drugs in the workplace, 162

Economic environment, 100, 101
Employee benefits, 156
Employee development, 149–152
Employee motivation, 7–11
Employee retention, 156, 157
Employment manual, 147–149
Employment practices liability insurance, 46
Equal employment opportunity, 162, 163

Equipment floater insurance, 41
Equipment purchases, 6
Errors and omissions insurance, 45, 46
Ethics, 7, 11–13, 137
Experienced-based insurance premiums, 43
Experience modification ratio (EMR), 47
External assessment, 97, 99–104
Extranets, 171

Financial analysis, 76–84
Financial management, 84–88
Financial statements, 73–76
    balance sheet, 73–75
    income statement, 75, 76
Firewalls, 173
Fixed assets, 75
Floating marine equipment insurance, 42
Functional organization, 24, 25
Functional strategies, 112, 113

General business environment,
    100–102, 104
General ledger system, 68
General partnership, 17–19
Gross profit, 75

Hardware, 169, 171, 172
Hierarchy of authority, 28
Human resources management, 136–165

Income statement, 75, 76
Indemnity agreement, 44, 51–53
Individual development plan, 150–152
Information management system, 168–170
Information security system, 173
Insurance, 36–48
Insurance agent, 37, 38, 45
Insurance certificate, 48–50
Insurance deductible, 36, 37, 39, 45
Insurance policy, 36
Internal assessment, 105–110
Internal audit, 108, 109
Internal customers, 8
Internal functional analysis, 109, 110
Internal marketing, 119, 127
Intranets, 171

Job analysis, 140
Job description, 140, 141
Job specifications, 140, 141
Joint venture, 17, 22, 23, 37

Key-person insurance, 48

Labor and materials bond, 56
Leadership, 7, 8, 137
Leadership development, 152, 153
Leading, 7
Leverage, 80
Liabilities, 68, 74
Liability insurance, 42–48
Limited liability company, 17, 22
Limited partnership, 17, 19, 20
Line of credit, 6, 87
Line and staff functions, 24
Liquidity analysis, 76, 77

Maintenance bond, 56, 57
Management functions, 7
Managerial maturity, 3
Market analysis, 122–126
Marketing, 117, 119–122
Marketing plan, 131–133
Marketing strategies, 126–129
Marketing tools, 129–131
Matrix organization, 26, 27
Mechanic's lien, 56
Mission statement, 95, 96

Net profit after taxes, 75
Net profit before taxes, 75
Net worth. *See* Owners' equity

Obligee, 48, 50
Occurrence-based policy, 44
Operations budget, 86
Organizational behavior, 8–11
Organizational design, 27–30, 140–142
Organizational structure, 23–27
Organization chart, 29, 30, 140
Organizing, 7
Overhead budget, 6, 87
Overhead rate, 87, 88
Owners' equity, 68, 74

Partnership, 17–19, 37
Payment bond. *See* Labor and materials bond
Penal sum, 48
People skills, 8
Percentage-of-completion accounting
    method, 72
Performance bond, 55
Performance management, 154–156

Performance measurement, 99
Planning, 7
Policies and operating procedures, 30, 31
Political/legal environment, 101
Pollution liability insurance, 46
Preventing violence in workplace, 163
Principal, 48, 50
Privacy rights, 163
Privately held corporations, 20
Process action teams, 186
Process improvement model, 182, 183
Professional liability insurance, 45
Profitability analysis, 78, 79
Project cost coding system, 71
Project indirect costs, 76
Property insurance, 38–42
Proprietorship, 17, 18
Publicly held corporations, 20

Quality council, 185, 186
Quality improvement cycle, 178, 179
Quality management board, 186
Quick ratio, 77, 78

Racial and ethnic discrimination, 163, 164
Railroad liability insurance, 45
Ratio analysis, 76
Recruiting, 143
Relevant market, 120
Religious protection, 164, 165
Retained earnings, 74, 86
Return on assets ratio, 78
Return on equity ratio, 79
Return on sales ratio, 79
Risk management, 34–36

Safety, 158–160
Sales, 117
Secondary accounting systems, 69, 71
Sexual harassment, 165
Short-term goals, 98
Signature authority policy, 31
Situation analysis, 92
Sociocultural environment, 102
Software, 169, 172, 173
Span of control, 28
Specific industry environment, 102–104
Staffing, 7, 142–147

Stockholders, 20, 21, 86
STP marketing concept, 120
Strategic management, 92
Strategic marketing, 119, 126, 127
Strategic objectives, 97, 98
Strategic planning, 91–113
Strategy evaluation, 92, 113, 114
Strategy formulation, 92, 110–113
Strategy implementation, 113
Subcontractor analysis, 102, 103
Subcontractor insurance, 48
Subrogation, 39, 40
Succession planning, 153, 154
Supplier analysis, 103
Supply bond, 57, 58
Surety bond, 48, 50, 51, 54–58
  bid bond, 54
  labor and materials bond, 56
  maintenance bond, 56, 57
  performance bond, 55
  supply bond, 57, 58
Surety bond producer. *See* Bonding agent
SWOT analysis, 99

Tactical marketing, 119, 127
Task interdependence, 28
Technical skills, 8
Technological environment, 102
Total quality management (TQM), 177–191
Transaction loans, 88
Transportation floater insurance, 41
Trend analysis, 81
Type S corporation, 17, 21, 22

Umbrella excess liability insurance, 44, 45
Underwriter, 59
Union relations, 157, 158

Value chain analysis, 105–108
Vision, 7, 96, 97

Warranty bond. *See* Maintenance bond
Wellness program, 158–160
Worker's compensation insurance, 46–48
Work flow, 28
Working capital, 77, 86